LIGHTING & ELECTRICITY

Other Publications:

LIGHTING & ELECTRICITY

TIME-LIFE BOOKS
ALEXANDRIA, VIRGINIA

Fix It Yourself was produced by
ST. REMY PRESS

MANAGING EDITOR	Kenneth Winchester
MANAGING ART DIRECTOR	Pierre Léveillé

Staff for *Lighting & Electricity*

Editor	Susan Bryan Reid
Art Director	Francine Lemieux
Research Editor	Elizabeth Cameron
Designer	Solange Pelland
Editorial Assistant	Michael Mouland
Contributing Writers	Beverley Bennett, Lydia Cassidy-Levasseur, Cathleen Farrell, Judee Ganten, Grant Loewen, Noel J. Meyer, Betty Palik, Jeremy Searle, Dianne Thomas
Contributing Illustrators	Gérard Mariscalchi, Jacques Proulx
Technical Illustrator	Robert Paquet
Cover	Robert Monté
Index	Christine M. Jacobs
Administrator	Denise Rainville
Coordinator	Michelle Turbide
Systems Manager	Shirley Grynspan
Systems Analyst	Simon Lapierre
Studio Director	Daniel Bazinet
Photographer	Maryo Proulx

Time-Life Books Inc. is a wholly owned subsidiary of
TIME INCORPORATED

FOUNDER	Henry R. Luce 1898-1967
Editor-in-Chief	Henry Anatole Grunwald
Chairman and Chief Executive Officer	J. Richard Munro
President and Chief Operating Officer	N. J. Nicholas Jr.
Chairman of the Executive Commitee	Ralph P. Davidson
Corporate Editor	Ray Cave
Group Vice President, Books	Kelso F. Sutton
Vice President, Books	George Artandi

TIME-LIFE BOOKS INC.

EDITOR	George Constable
Executive Editor	Ellen Phillips
Director of Design	Louis Klein
Director of Editorial Resources	Phyllis K. Wise
Editorial Board	Russell B. Adams Jr., Thomas H. Flaherty, Lee Hassig, Donia Ann Steele, Rosalind Stubenberg, Kit van Tulleken, Henry Woodhead
Director of Photography and Research	John Conrad Weiser
PRESIDENT	Christopher T. Linen
Chief Operating Officer	John M. Fahey Jr.
Senior Vice Presidents	James L. Mercer, Leopoldo Toralballa
Vice Presidents	Stephen L. Bair, Ralph J. Cuomo, Neal Goff, Stephen L. Goldstein, Juanita T. James, Hallett Johnson III, Robert H. Smith, Paul R. Stewart
Director of Production Services	Robert J. Passantino

Editorial Operations

Copy Chief	Diane Ullius
Editorial Operations	Caroline A. Boubin
Production	Celia Beattie
Quality Control	James J. Cox (director)
Library	Louise D. Forstall
Correspondents	Elisabeth Kraemer-Singh (Bonn); Maria Vincenza Aloisi (Paris); Ann Natanson (Rome).

THE CONSULTANTS

Consulting Editor **David L. Harrison** is Managing Editor of Bibliographics Inc. in Alexandria, Virginia. He served as an editor of several Time-Life Books do-it-yourself series, including *Home Repair and Improvement, The Encyclopedia of Gardening* and *The Art of Sewing.*

Mark M. Steele, a professional home inspector in the Washington, D.C. area, is an editor of home improvement articles and books.

Joseph A. Tedesco was Associate Editor of the National Fire Protection Association 1984 National Electrical Code Handbook. He advises the electrical industry on the National Electrical Code.

Klaus Bremer, special consultant for Canada, is a master electrician. He operates Bremer Electric, an electrical contracting firm in Montreal.

Library of Congress Cataloguing in Publication Data
Lighting & electricity
  (Fix it yourself)
  Includes index.
  1. Electric lighting—Amateurs' manuals. 2. Electric lamps—Amateurs' manuals. 3. Electric wiring—Amateurs' manuals. I. Time-Life Books. II. Title: Lighting and electricity. III. Series.

TK9921. L54 1987	621.32'2	87-6476

ISBN 0-8094-6248-6
ISBN 0-8094-6249-4   (lib. bdg.)

For information about any Time-Life book,
please write:
Reader Information
541 North Fairbanks Court
Chicago, Illinois 60611

CONTENTS

HOW TO USE THIS BOOK

Lighting & Electricity is divided into three sections. The Emergency Guide on pages 8-13 provides information that can be indispensable, even lifesaving, in the event of a household emergency. Take the time to study this section *before* you need the important advice it contains.

The Repairs section—the heart of the book—is a system for troubleshooting and repairing switches, outlets, lamps and lighting fixtures. Pictured below are four sample pages from the chapter on floor and table lamps, with captions describing the various features of the book and how they work. If the bulb in your lamp flickers, for example, the Troubleshooting Guide will offer a number of possible causes. If the problem is a faulty socket, you will be directed to page 30 for detailed, step-by-step directions for removing and replacing it.

Each job has been rated by degree of difficulty and the average time it will take for a do-it-yourselfer to complete. Keep in mind that this rating is only a suggestion. Before deciding whether you should attempt a repair, first read all the instructions carefully. Then be guided by your own confidence, and the tools and time available to you. For more complex or time-consuming repairs, such as extending an existing circuit or installing a ceiling box and chandelier, you may wish to call for

---

Introductory text
Describes proper use and care of the electrical part or fixture, most common breakdowns and basic safety precautions.

"Exploded" and cutaway diagrams
Locate and describe the various parts of the fixture or device.

Troubleshooting Guide
To use this chart, locate the symptom that most closely resembles your electrical problem, review the possible causes in column 2, then follow the recommended procedures in column 3. Simple fixes may be explained on the chart; in most cases you will be directed to an illustrated, step-by-step repair sequence.

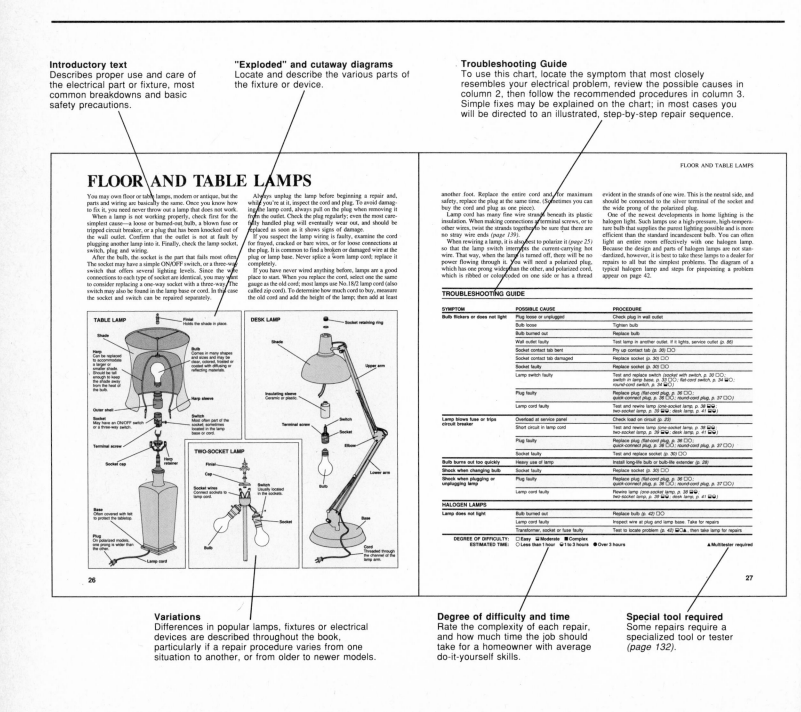

Variations
Differences in popular lamps, fixtures or electrical devices are described throughout the book, particularly if a repair procedure varies from one situation to another, or from older to newer models.

Degree of difficulty and time
Rate the complexity of each repair, and how much time the job should take for a homeowner with average do-it-yourself skills.

Special tool required
Some repairs require a specialized tool or tester *(page 132)*.

professional service. You will still have saved time and money by diagnosing the problem yourself.

Most of the repairs in *Lighting & Electricity* can be made with a screwdriver, a continuity tester, a voltage tester, diagonal-cutting pliers and a multipurpose tool. You may need a stepladder (preferably wooden) to work on hard-to-reach fixtures. Basic electrical tools—and the proper way to use them—are presented in the Tools & Techniques section starting on page 132. If you are a novice when it comes to home repair, read this section and the chapter called Electricity in the Home *(page 14)* in preparation for a major job.

Working with electricity is easy and safe if you work logically and follow the safety precautions. Before beginning any electrical repair, turn off power to the circuit by removing the fuse or tripping the circuit breaker at the house's main service panel. Then use a voltage tester *(page 134)* to confirm that the power is off. If the tester glows, indicating that there is still current in the wires, return to the panel to find the correct fuse or circuit breaker. Continue testing until you are sure the power is off. Take special precautions when working in wet conditions *(page 16)*. Most important, follow all safety tips and **Caution** warnings throughout the book.

Name of repair
You will be referred by the Troubleshooting Guide to the first page of a specific repair job.

Step-by-step procedures
Follow the numbered repair sequence carefully. Depending on the result of each step, you may be directed to a later step, or to another part of the book, to complete the repair.

Insets
Illustrate variations of popular models, and provide close-up views of specific steps.

Tools and techniques
When a tool or method is required for a job, it is described within the step-by-step repair. General information on using electrical testers is covered in the Tools & Techniques section *(page 132)*.

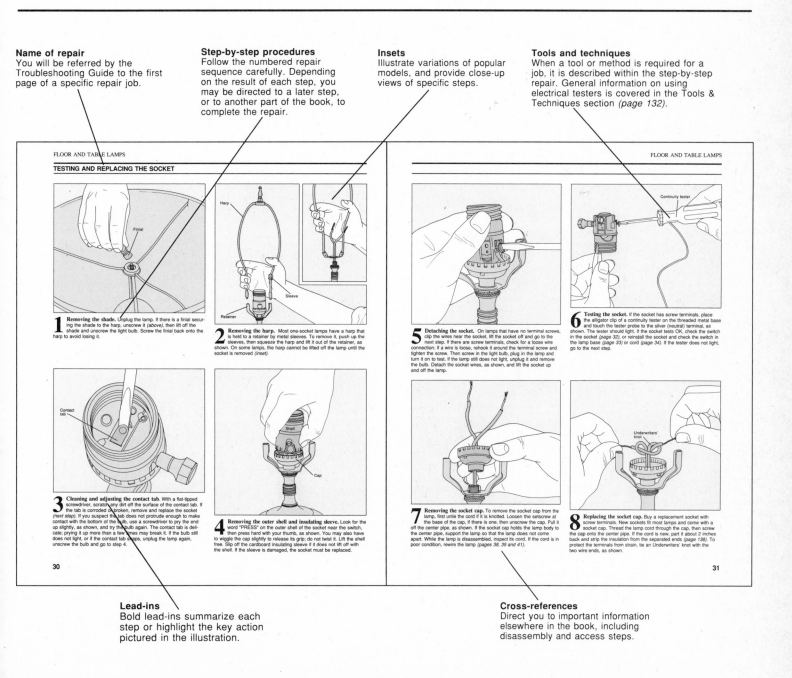

Lead-ins
Bold lead-ins summarize each step or highlight the key action pictured in the illustration.

Cross-references
Direct you to important information elsewhere in the book, including disassembly and access steps.

7

EMERGENCY GUIDE

Preventing electrical problems. Your home's wiring is mostly invisible—safely concealed behind walls and cover plates. More than any other system in your house, it is strictly regulated by codes and standards to protect against fire and shock. Properly inspected and maintained, the system will rarely pose a hazard. Repairing electrical fixtures and wiring is equally safe, as long as you remember two basic rules: Always turn off the power before working on your wiring *(page 18)*, then confirm that it is off by using a voltage tester *(page 134)*. Never touch any screw terminals, bare wire ends, sockets or metal boxes until you have confirmed that the power is off.

Most emergencies occur at the points where electrical equipment is handled: in switches, outlets, fixtures, lamps, appliances and power tools that are subject to daily wear and tear. Such hazards usually announce themselves in the form of a frayed cord or cracked plug long before they threaten life or property. Dealing with minor problems immediately will prevent them from becoming major emergencies.

Fire is the most dangerous of all electrical emergencies. When the normal flow of current is impeded by a broken wire or faulty connection, heat can build up. Periodically inspect your wiring, appliances and lighting fixtures and replace any worn or damaged parts. When electrical fires occur, they should be treated with extra caution *(page 11)*. First, turn off power to the circuit, then use only a dry-chemical fire extinguisher. Fire departments welcome requests to inspect possible fire hazards and to investigate even the suspicion of fire.

Water can turn a small electrical fault into a deadly hazard by making your body a convenient path for electrical current. Do not handle electrical devices, even switches or power cords, in wet conditions. If you must rescue someone stuck to a live current, do not touch him. Turn off the power at the service panel or use a wooden chair or broom handle to knock him free *(page 10)*.

Before emergency situations arise, read the Troubleshooting Guide on page 9, which places emergency procedures at your fingertips. It lists quick-action steps to take, and refers you to procedures on pages 10-13 for more detailed instructions. Also review the list of safety tips at right and the introduction to electricity starting on page 14, and familiarize yourself with the Tools & Techniques section *(page 130)*.

When in doubt about the safety of your wiring or your ability to handle an emergency, don't hesitate to call for help. Post the telephone numbers for the police, fire department and power company near the telephone. Even in non-emergency situations, an inspector from your power company or fire department can answer questions concerning the condition and proper use of your electrical system.

SAFETY TIPS

1. Before attempting any repair in this book, read the entire procedure. Familiarize yourself with the specific safety information presented in each chapter.

2. Read the introductory chapter Electricity in the Home *(page 14)* to better understand how to live and work safely with electricity.

3. Map and label your home's electrical circuits *(page 20)* to identify when the electricity is live and dangerous or switched off and safe.

4. Never work on service panel wiring. Entrance wires may remain live even when the main breaker or fuse is off.

5. Never work with electricity in wet conditions and do not use wet electrical cords, tools or appliances.

6. Guard against shock when working with electric tools or appliances in damp conditions or outdoors *(page 16)*. Use a GFCI-protected outlet.

7. Do not touch a metal faucet, pipe or appliance, or any other ground, when working with electricity.

8. Keep protective equipment handy: safety goggles for overhead work, heavy rubber work gloves, rubber-soled shoes for working outdoors or in damp conditions.

9. Before working on a circuit, switch off the power at the service panel *(page 18)*. Leave a note on the panel so that no one switches the circuit back on while you are working.

10. Light your service panel and work area well. Keep a flashlight near the service panel.

11. Use only replacement parts of the same specifications as the original, or upgraded according to federal and local codes. Look for the UL (Underwriters Laboratories) or CSA (Canadian Standards Association) logo on new parts.

12. Do not splice electrical cords or remove the grounding prong from a three-prong plug.

13. Do not use extension cords as permanent wiring. Never tack them down.

14. If you replace a metal pipe with plastic, check to see if the original pipe was part of the house's electrical grounding system. If so, have an electrician install a grounding jumper.

15. Let a lamp or fixture cool before working on it. Never wipe a hot bulb with a wet cloth.

16. Use only materials permitted by local electrical codes. If in doubt, check with local authorities before working on your system.

17. If in doubt about the safety of any electrical repair, have it inspected by your power company or an electrician.

18. Post emergency, utility company and repair service numbers near the telephone.

19. Install smoke detectors and fire extinguishers in your home *(page 12)*.

20. Do not allow children to play with electrical cords, lamps or appliances. Cover accessible outlets with plastic safety caps.

TROUBLESHOOTING GUIDE

PROBLEM	PROCEDURE
Appliance, lamp or power tool gives off sparks or shocks user	Unplug cord *(p. 11)* without touching appliance or tool, or turn off power at service panel *(p. 10)*
Cord or plug sparking or hot to the touch	Turn off power at service panel *(p. 10)*, then unplug cord using a towel *(p. 11)*
Switch sparking or hot to the touch	Turn off switch using a wooden spoon or broom handle *(p. 12)*, or turn off power at service panel *(p. 10)*
Fixture sparking or hot to the touch	Turn off wall switch, then turn off power at service panel *(p. 18)*
Arcs and sparks at service panel	Do not touch service panel; call power company to have power turned off or call an electrician to repair panel
Small appliance falls in sink or bathtub	Do not touch appliance or any plumbing fixture. If you are dry, pull out cord or turn off power to appliance at service panel *(p. 18)*
Large appliance or outlet wet or submerged	Do not enter room. If conditions around service panel are dry, turn off power at service panel *(p. 10)*; otherwise, leave house and call power company
Basement or room flooded	Call electric company to disconnect power to your home
Fire in electrical outlet, switch, fixture, appliance or cord	Call fire department
	Use fire extinguisher rated for electrical fires *(p. 11)*
	Turn off power at service panel *(p. 10)*, then unplug cord *(p. 11)*
	If flames or smoldering continue, leave house and wait for fire department
Power outage	Turn off all appliances with motors or heating elements, including furnace, air conditioner, heater, washer and dryer to prevent overloading system when power is restored
	Check service panel. If main circuit breaker has tripped or main fuses have blown, call an electrician or power company to inspect system
	Have emergency supplies on hand, including a small space heater and lantern, flashlight or candles. A portable generator can provide a limited amount of emergency power *(p. 13)*
	Leave several lights on so that you will know when power has been restored
Power line fallen in yard	Call power company, police or fire department
	Treat every downed wire as a live wire. Stay far away from fallen power line and anything it touches, including fences and trees
Power line fallen on car	Stay in car until help arrives, and warn others to stay away
	Have someone call power company, police or fire department
	A car's rubber tires insulate it against accidental grounding; do not attempt to jump clear of car and do not touch any metal parts
Person trapped under downed power line	Do not attempt to rescue; call power company, police or fire department
Child with utensil in outlet or toaster	Push child away with wooden spoon, broom handle or chair *(p. 10)*
	Treat child for injuries or call for help *(p. 10)*
Injury due to electrical shock	Check to see if victim is breathing and has a pulse. If not, begin artificial resuscitation or cardiopulmonary resuscitation (CPR) if you are qualified to do so. Otherwise, place victim in recovery position *(p. 10)* and call for help
Lightning storm	Unplug electronic devices or use a surge suppressor to protect them. Be careful near windows, doors, fireplaces, radiators, stoves, sinks and pipes

TURNING OFF POWER AT THE SERVICE PANEL

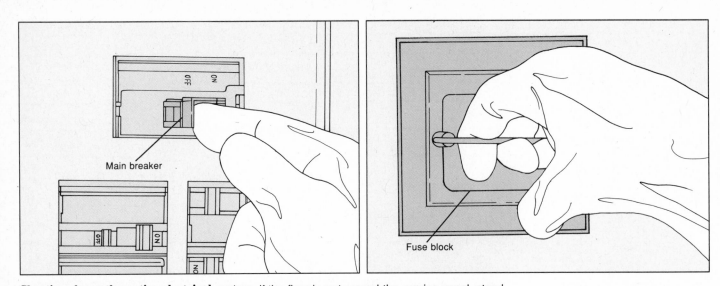

Shutting down the entire electrical system. If the floor is wet around the service panel, stand on a dry board or rubber mat, or wear rubber boots. Wear heavy rubber gloves or use a wooden broom handle. Work with one hand only, to protect your body from becoming a path for electrical current, and keep the other hand in your pocket or behind your back. At a circuit breaker panel *(above, left)*, flip off the main breaker. As an added precaution, use your knuckle; any shock will jerk your hand away from the panel. At a fuse panel *(above, right)*, remove the main fuse block by gripping its metal handle and pulling it from the box. Remove all main fuse blocks if there is more than one. Some fuse panels have a shutoff lever instead of a fuse block.

RESCUING A VICTIM OF ELECTRICAL SHOCK

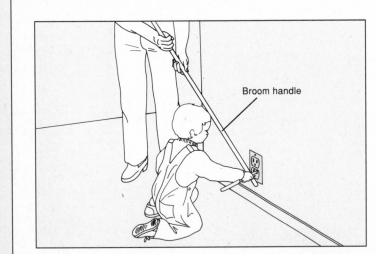

Freeing someone from a live current. Usually a person who accidentally contacts live current will be thrown back from the source. But if the victim is stuck (muscles may contract involuntarily around a wire or appliance), do not touch him. Disconnect the appliance or lamp by pulling its plug *(page 11)*, or shut off power to a switch or outlet at the main service panel. If the power cannot be cut immediately, use a wooden broom handle or chair to knock the person free, as shown.

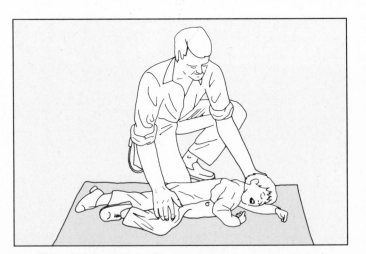

Handling victims of electrical shock. If the victim is unconscious or burned, call for help immediately. Check for breathing and heartbeat. If the victim is breathing and has not received back or neck injuries, place him in the recovery position, as shown. Tilt the head back with the face to one side and the tongue forward to maintain an open airway. If there is no sign of breathing or heartbeat, give mouth-to-mouth resuscitation or cardiopulmonary resuscitation only if you are qualified to do so.

ELECTRICAL FIRE

Class ABC or BC extinguisher

Using a fire extinguisher. Call the fire department as soon as possible, even if the fire has been put out. If there are flames or smoke coming from the walls or ceiling, leave the house to call for help. To snuff a small, accessible fire in an appliance, outlet, switch or light fixture, use a dry-chemical fire extinguisher rated ABC or BC. Stand near an exit, 6 to 10 feet from the fire. Pull the lock pin out of the extinguisher handle and, holding the extinguisher upright, aim the nozzle at the base of the flames. Squeeze the two levers of the handle together, spraying in a quick side-to-side motion. You may also have to turn off power at the service panel *(page 10)* to remove the source of heat at the fire. Keep spraying until the fire is completely extinguished. Watch carefully for "flashback," or rekindling, and be prepared to spray again. Find the cause of the fire and replace damaged wiring and devices before restoring the circuit to use.

HOT CORD, PLUG OR APPLIANCE

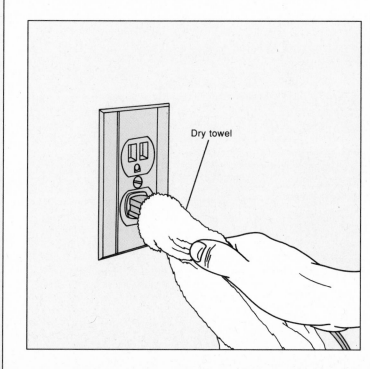

Dry towel

Pulling the power cord. Caution: If the floor or counter is wet, or the outlet itself is sparking or burning, do not touch the cord, lamp or appliance. Instead, turn off power to the circuit at the service panel *(page 18)*. If the lamp or appliance sparks, shocks you, feels hot or is burning, disconnect the plug. Protect your hand with a thick, dry towel or a heavy work glove. Without touching the outlet, grasp the cord with one hand several inches from the plug, as shown, and pull it out. Locate and repair the problem before using the device again.

SPARKING SWITCHES, OUTLETS AND FIXTURES

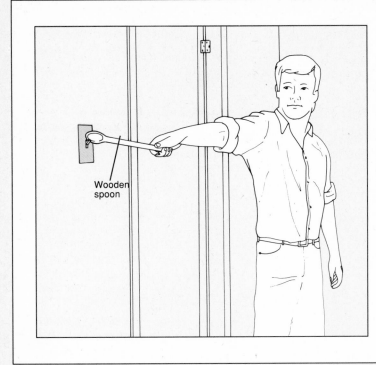

Switching off electrical hazards. If a switch, outlet or fixture makes the snapping or crackling sounds of electrical arcing, or if visible sparks, smoke or flames appear, immediately turn off power to that circuit at the service panel *(page 18)*. If in doubt about which fuse or breaker controls the circuit, turn off all electricity *(page 10)*. Find and repair the fault before restoring power to the affected circuit *(page 104)*.

Never touch a burning or sparking switch to turn it off. Stand away from it and flip off the toggle with a wooden spoon, as shown.

SAFETY ACCESSORIES

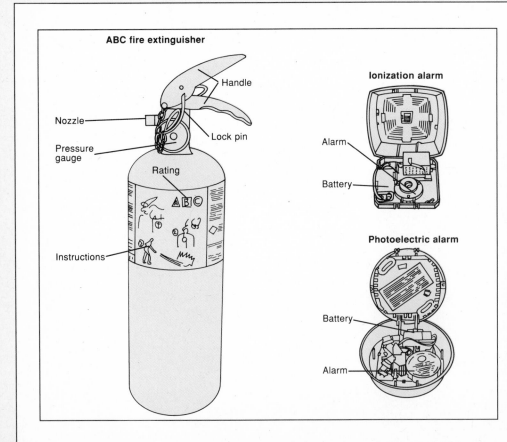

Necessary tools for fire prevention. Best for use in the home is a multipurpose dry-chemical extinguisher rated ABC *(far left)*. The A and B ratings indicate effectiveness against wood or upholstery fires and flammable liquid fires. Electrical fires require a C rating. A convenient size holds a pressurized cargo of 2 1/2 to 7 pounds. Check the pressure gauge monthly; after any discharge or loss of pressure, have the tank recharged professionally or buy a new extinguisher. Mount extinguishers, using the wall bracket provided, near doors to the kitchen, garage and basement.

Install at least one smoke alarm in a central hallway on every floor of the house, near the kitchen, bedrooms and head of the stairs, as well as in the garage and basement. Ionization alarms *(left, top)*, which sense atomic particles, respond quickly to hot fires with little smoke, but they tend to set off annoying false alarms in the presence of normal cooking fumes. Photoelectric alarms *(left, bottom)* "see" smoke molecules: they respond best to the smoldering typical of cooking, appliance and upholstery fires.

COPING WITH POWER FAILURE

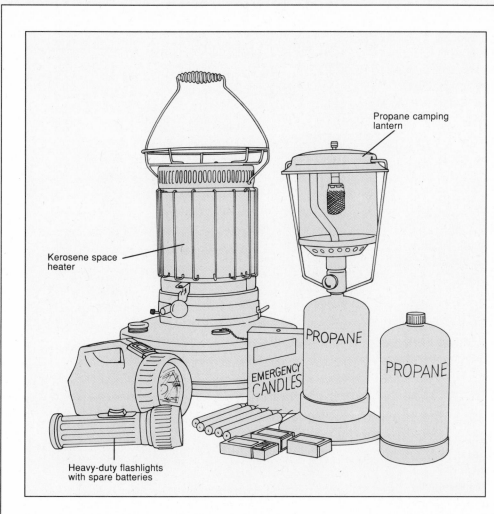

Propane camping lantern

Kerosene space heater

EMERGENCY CANDLES

PROPANE

PROPANE

Heavy-duty flashlights with spare batteries

Emergency supplies. In case your power fails for several hours, have alternate sources of light available. Candles are a reliable source of light and provide a surprising amount of warmth in small, well-insulated living spaces. Keep candles, matches and a flashlight in a familiar, accessible location. A propane lantern will provide bright, long-lasting light but should be used with caution: when fuel-fired heaters are used indoors, always open a door or window slightly for proper ventilation. A kerosene heater rated for indoor use is a safe source of heat provided the manufacturer's instructions regarding fuel, ventilation and operation are strictly followed.

To prepare for power restoration, reduce the load on your electrical system by turning off or unplugging all heating and motorized equipment. Open your refrigerator or freezer as little as possible; spoilage will not likely occur within 24 hours.

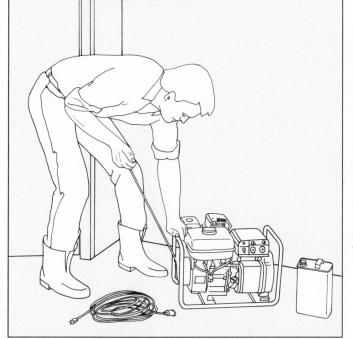

A portable electric generator. A gas-powered generator can be purchased or rented to deal with longer power outages. They must be operated in a dry, sheltered area with direct ventilation to the outside, such as a garage or porch. Only appliances that plug into extension cords should be attached to the generator (a permanent hookup to the house wiring must be left to a professional). A 220-watt unit, as shown—large enough to operate a refrigerator or freezer, a TV and several lamps—will weigh about 50 pounds. Calculate the load you need to run *(page 23)* in order to choose the right size. If you are running cooling compressors or motorized tools that require an initial surge of voltage to start up, the generator will have to be larger or have surge capacity.

ELECTRICITY IN THE HOME

The first time you lift off a switch cover plate or disassemble a table lamp, you may be overwhelmed by the tangle of wires ("This looks much too complicated!"). Or you may be daunted by the notion that electrical repair is dangerous. ("Electricity can kill, can't it?"). But one of the best things about electricity is that it follows a relentless logic. If you have a basic understanding of your home's wiring, work methodically and take reasonable precautions, you can tackle any repair in this book with confidence.

Electrical current flows in a continuous path, or circuit, from a power source through various switches, fixtures and appliances, and then back to the source. In a household electrical system, the source is the main service panel and the path is your home's wiring. Electricity always takes the path of least resistance. Copper and aluminum are good conductors of electricity and are therefore used to carry current; plastic, rubber, glass and porcelain are non-conductive and are used as insulators for electrical equipment.

The power company delivers electricity to your home via three large overhead or underground wires that arrive at the service entrance. From there, electricity passes through a meter into the main service panel, where it is divided into branch circuits and distributed throughout the house. Two of the three service wires are hot and carry 120 volts each, providing power to operate 120-volt and 240-volt appliances.

To prevent damage to the wiring and guard against fire and shock, each circuit is protected by a fuse or circuit breaker. If there is an overload or break in the circuit, this device will instantly stop the flow of current. The main circuit that brings electric current into the house is also equipped with a fuse or circuit breaker that will turn off all current at the panel.

Normally, cables that carry 120-volt current contain a black wire, a white wire and a bare copper grounding wire. From the service panel, the black (hot) wire delivers 120 volts of current "under pressure" to light a bulb or run an appliance. The white wire carries the electrons at close to zero volts back to the service panel. In an analogy to home plumbing, the black wire is equivalent to the supply pipe and the white wire serves as the drainpipe.

The bare copper grounding wire (or insulated green wire) in each circuit safeguards the system by providing a second path for electricity to return to the service panel. A main grounding wire runs from the service panel to a metal water pipe or a metal grounding rod that is buried in the earth.

Electrical codes
The NEC (National Electrical Code) sets strict safety standards for electrical procedures and equipment. Before undertaking any electrical work, consult your local building authority to find out about national and local code requirements.

Doorbell
Power is supplied via a transformer, which steps down 120-volt household current to 6-24 volts.

Water pipe
Code now requires that the grounding connection to the water pipe be supplemented by a second grounding connection, commonly a metal rod driven at least 8 feet into the earth.

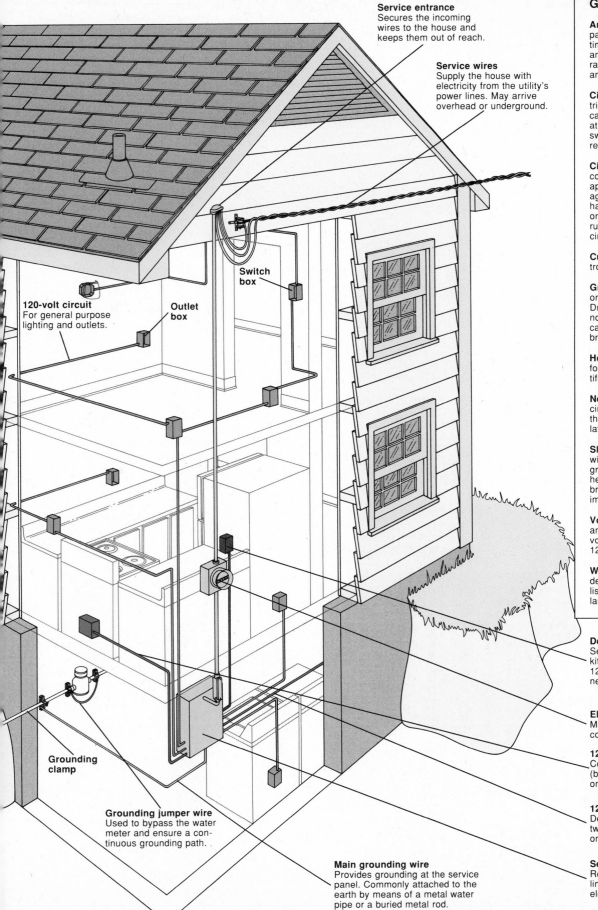

Service entrance
Secures the incoming wires to the house and keeps them out of reach.

Service wires
Supply the house with electricity from the utility's power lines. May arrive overhead or underground.

Switch box

120-volt circuit
For general purpose lighting and outlets.

Outlet box

Grounding clamp

Grounding jumper wire
Used to bypass the water meter and ensure a continuous grounding path.

Main grounding wire
Provides grounding at the service panel. Commonly attached to the earth by means of a metal water pipe or a buried metal rod.

GLOSSARY

Amperes: The amount of current passing a given point at a given time. Each electrical device has an amperage rating and each circuit is rated for the total number of amperes it can safely deliver.

Circuit: A continuous path for electrical current. In a household electrical system, a branch circuit begins at the service panel, runs to various switches, outlets and fixtures and returns to the service panel.

Circuit overload: Occurs when the combination of lights, tools and appliances is drawing more amperage than the circuit is designed to handle. Normally, the fuse will blow or the ciruit breaker will trip, interrupting the flow of electricity to the circuit.

Current: The movement of electrons, measured in amperes.

Grounding wire: The bare copper or green insulated wire in a cable. Drains off current that escapes its normal path to the service panel, causing a fuse to blow or a circuit breaker to trip.

Hot wire: A wire that carries current forward from the source. Often identified by black (or red) insulation.

Neutral wire: Used to complete the circuit by carrying current back to the source. Identified by white insulation.

Short circuit: When an exposed hot wire touches a neutral wire or a grounded metal box, the circuit will heat up suddenly. The fuse or breaker will shut off the power immediately.

Volts: The strength, or pressure, of an electrical current, measured in volts. Household circuits are usually 120 volts, 240 volts or low-voltage.

Watts: The rate at which electrical devices consume energy. Usually listed on a sticker or plate on the lamp or appliance.

Dedicated 120/240-volt circuit
Serves heavy-duty appliances in kitchens and workshops. Contains two 120-volt hot wires (black and red), one neutral and one grounding wire.

Electric meter
Measures the total electricity consumed in kilowatt-hours.

120/240-volt circuit
Contains two 120-volt hot wires (black and red), one neutral and one grounding wire.

120/240-volt circuit
Dedicated to clothes dryer. Contains two 120-volt hot wires (black and red), one neutral and one grounding wire.

Service panel
Receives three lines: two 120-volt hot lines and one neutral line; distributes electricity throughout the house.

WORKING SAFELY WITH ELECTRICITY

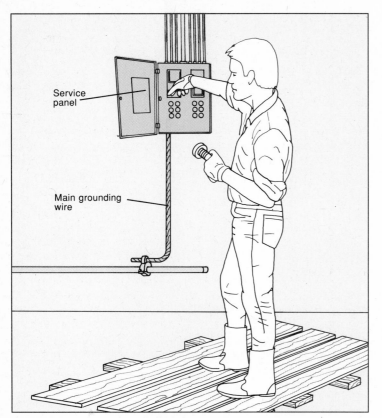

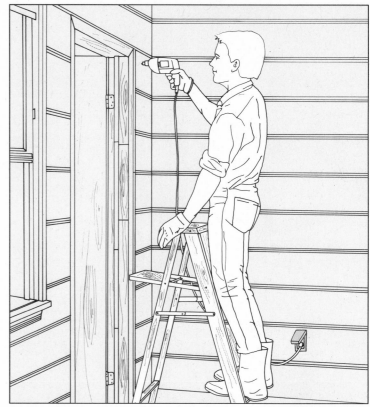

Safety at the service panel. When working at the service panel, even to change a fuse or reset a circuit breaker *(page 18)*, be sure to take basic safety precautions. Dry any water on the floor. To protect your body from making a circuit to the ground, stand on boards or wear dry rubber boots and heavy rubber gloves. Work with one hand only, keeping the other in your pocket or behind your back to avoid touching anything metal. Have a flashlight at a convenient spot near the service panel so that you do not have to change a fuse in the dark. Do not touch a service panel that is sparking, blackened or rusted; call an electrician. Do not remove the panel cover to expose the service cables; even if you have turned off the main breaker or pulled the main fuse block, parts of the box remain charged with current.

Safety outdoors. Code now requires that new outdoor outlets be protected by a GFCI (ground-fault circuit interrupter). This safety device is especially important in damp locations, where electrical shocks can be severe. When working outdoors with a power tool, be sure that it is properly grounded. As an added safety precaution, stand on a wooden plank or rubber mat and use a wooden ladder and heavy rubber gloves. Be careful not to touch overhead power lines when working on your roof or siding. Call the power company to locate any underground power lines before digging in the yard.

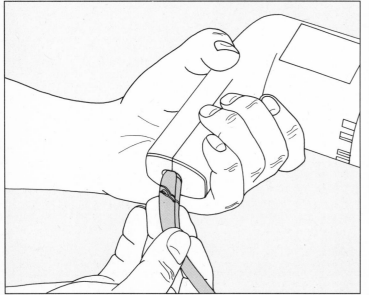

Safety and power tools. Inspect power tools regularly for signs of wear. Look for corroded, loose or bent plug prongs and examine the cord for cracked or frayed insulation; some tools have a strain-relief sleeve to protect the cord from wear at the point where it is joined to the tool. To prevent electrical shock, power tools should have a three-prong plug or double insulation—a plastic housing that isolates metal parts from contact with the hand. Do not touch a faucet, water pipe or another grounded appliance while using a power tool. If a tool sparks, shocks or becomes hot to the touch, it must be repaired or replaced. Never carry a power tool by its cord. All pliers, screwdrivers and stripping tools used for electrical work should have insulated handles, or handles wrapped in electrical tape.

LIVING SAFELY WITH ELECTRICITY

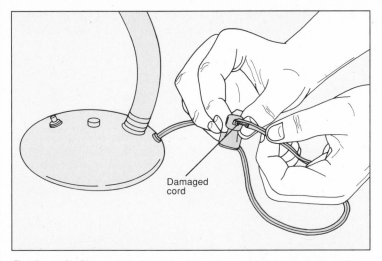

Cords and plugs. Inspect cords and plugs regularly and replace them if they appear damaged or heat up when the appliance is in use. Cords tend to crack or fray at the plug and at the appliance. Any cracking, tearing or rubbing of the insulation will eventually expose bare wire and create a fire or shock hazard. Keep cords away from heat and water, which can damage the insulation, and do not run them under rugs, where they are subject to friction. Plugs with removable insulating discs are unsafe, especially if the disc is missing and the wire ends are exposed. When removing a plug from an outlet, pull on the plug, not the cord. Never break off the third prong on a grounded plug to adapt it to a two-slot outlet. Instead, replace the two-slot outlet with a three-slot, grounded outlet *(page 98)*.

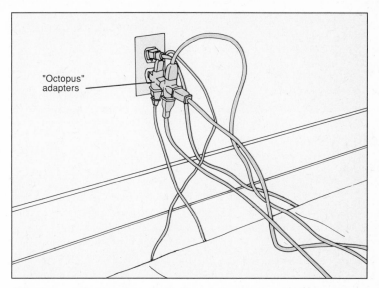

Adapters and extension cords. Do not use "octopus" connections; plugging too many cords into one outlet could overload the circuit. If a plug is loose or its prongs exposed, the poor connection could produce heat and sparking. Extension cords are not designed to take the place of permanent wiring; use them to bring power into an area only temporarily. Never attach extension cords with tacks or pins. Permanent use of extension cords and octopus adapters is an indication that your home's wiring is inadequate and should be updated.

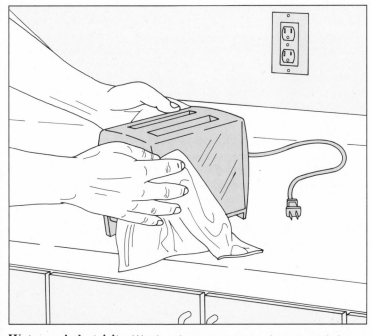

Water and electricity. Wet hands can create an alternate path for electrical current. Unplug a small appliance before cleaning it with a wet cloth, then dry your hands—and the appliance—thoroughly before plugging it back in. Radios, hair dryers and shavers are particular hazards in bathrooms. Areas of the home where there is plumbing or dampness should be protected by a ground-fault circuit interrupter *(page 25)*, which will shut down the circuit if leakage is detected.

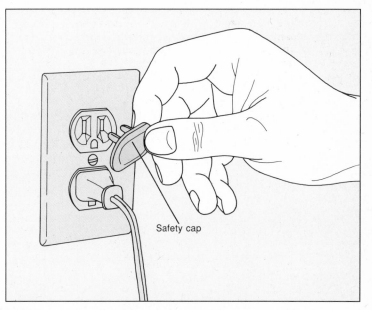

Children and electricity. Teach children to treat electricity with respect. Instruct them not to play with cords and wires and never to poke things into appliances or outlets. To protect curious fingers from a nasty shock, cover all unused outlets with plastic safety caps that fit tightly in the slots.

CIRCUIT-BREAKER PANEL

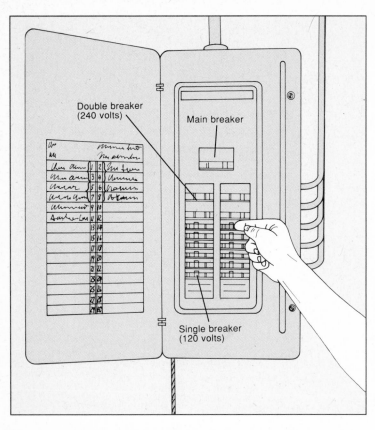

Double breaker (240 volts)

Main breaker

Single breaker (120 volts)

Resetting a tripped circuit breaker. Each circuit served by the circuit-breaker panel is controlled by a switch that will flip off automatically if there is a short circuit or overload. When tripped, some breakers flip to the OFF position; others flip to a middle position *(below)*. To reset a breaker that is in the tripped position, wait a minute for the heater strip to cool, then push the toggle to OFF, then to ON *(left)*. If the breaker snaps off again immediately, disconnect one or two appliances, or inspect the circuit for a short *(page 104)*. If the problem is not a short circuit or an overload, have an electrician inspect the panel.

Before working on a switch, outlet or fixture, turn off the circuit that supplies power to the device by flipping the breaker toggle to the full OFF position. Work only in dry conditions, use one hand, and do not touch any metal parts with the other hand. If the circuit breakers in the panel box are not correctly labelled, follow the instructions on page 20 to map your home's wiring. In an emergency, the entire system can be shut down by flipping off the main breaker *(page 10)*.

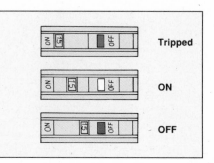

Tripped

ON

OFF

FUSE PANEL

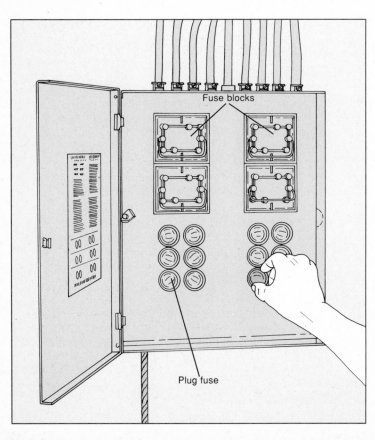

Fuse blocks

Plug fuse

Removing and replacing a fuse. Fuse panels can be found in houses built before 1960. When the strip of metal inside a fuse is melted by excess current, the fuse must be replaced *(below)*. A complete break in the metal strip indicates a circuit overload; move one or two appliances to another circuit before replacing the fuse. A discolored fuse points to a short circuit. Inspect the appliances, then the electrical boxes along the circuit *(page 104)* before replacing the fuse.

Turn off or unplug the appliances on the circuit before removing a plug fuse. Grasp the fuse by its insulated rim only and unscrew it *(left)*; use one hand and do not touch any metal parts. Replace a blown fuse with one of the same amperage. Inspect the fuse panel every six months and check that the fuses are tight. A loose fuse can overheat and cause arcing at the panel.

Before working on a switch, outlet or fixture, remove the fuse that supplies power to that circuit *(left)*. The 240-volt outlets are controlled by cartridge fuses housed in fuse blocks or in a separate panel. Work only in dry conditions, use one hand, and do not touch any metal parts with the other hand. If the fuses in the panel box are not correctly labelled, follow the instructions on page 20 to identify the circuits. In an emergency, turn off power to all circuits by pulling the main fuse blocks *(page 10)*.

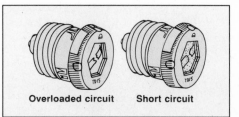

Overloaded circuit   Short circuit

PLUG AND CARTRIDGE FUSES

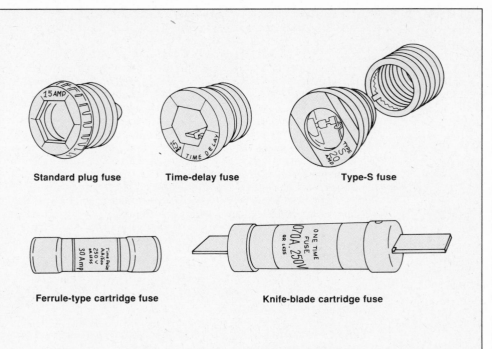

Standard plug fuse

Time-delay fuse

Type-S fuse

Ferrule-type cartridge fuse

Knife-blade cartridge fuse

Standard plug fuse. Plug fuses come in 15-, 20- and 30-ampere versions; their amperage must match the gauge of the wire in the circuit. Never replace a blown fuse with one of higher amperage.

Time-delay fuse. The metal strip of a time-delay fuse will withstand the momentary power surge created when an appliance motor starts up, but will blow if there is a sustained overload or a short circuit.

Type-S fuse. This fuse is designed to fit into an adapter that screws into the panel. The adapter accepts only a Type-S fuse of matching amperage, guarding against accidental installation of a higher amperage fuse.

Ferrule-type cartridge fuse. Found in fuse blocks in the main, or in a separate panel, this fuse protects a separate circuit for a large appliance. It is rated up to 60 amperes.

Knife-blade cartridge fuse. This cartridge fuse—rated over 60 amperes—is used to protect the house electrical system.

TESTING AND REPLACING CARTRIDGE FUSES

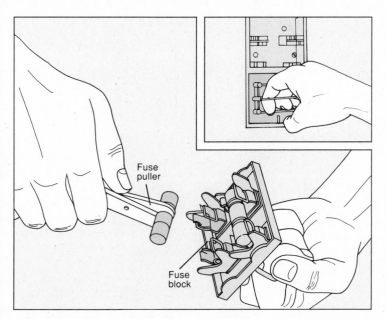

Fuse puller

Fuse block

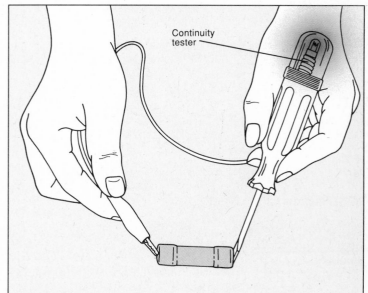

Continuity tester

1 Removing a cartridge fuse. If an appliance protected by a cartridge fuse fails to work, turn off or unplug the appliance and check the fuse. If the fuse is housed in a block in the main service panel, grasp the handle of the fuse block firmly with one hand and pull it out *(inset)*. Release the fuse from the spring clips with a fuse puller, as shown, or by hand. Do not touch the metal ends, which may be hot. If the cartridge fuse is housed in a separate panel box, use the plastic fuse puller to remove it. Work only in dry conditions, use one hand, and do not touch any metal parts.

2 Testing the fuse. Since a cartridge fuse shows no visible sign of damage, you will need a tester to determine if it is blown. Touch the alligator clip of the continuity tester to the metal cap at one end of the fuse and touch the tester probe to other end *(above)*. The tester will light if the fuse is good. If the tester doesn't light, replace the fuse with one of the same amperage. Push the new cartridge fuse against the spring clips until it snaps into place. If the new fuse blows when the appliance is turned on, there is probably a short circuit in the appliance.

MAPPING YOUR HOME'S WIRING

Tracing each circuit. Labels on service panels are often incorrect or out-of-date. Before performing any work on your home's electrical system, it is important to locate and correctly label the branch circuits served by the service panel. Begin by sketching a map of each floor of the house. (You can use the grids provided on pages 21 and 22.) Then walk around each floor, sketching in the outlets, switches, fixtures and major appliances in each room. Next, turn off all switches and appliances. At the service panel, post a new label, which you can obtain from the power company or an electrical supplier. Clearly number each fuse or circuit breaker in the service panel, then turn off power to circuit Number 1, using one hand to remove the fuse or trip the circuit breaker *(page 18)*. To identify the devices on the circuit, find the switches, outlets and fixtures that are no longer receiving power. Flip on switches and small appliances and plug a reliable lamp into the upper and lower receptacle of each outlet; those that do not work are controlled by circuit Number 1. (Note that a circuit can serve more than one room or floor.) On your map, write the circuit number beside each switch, outlet or fixture served by the first circuit, then return to the service panel and turn on the power to that circuit. Turn off power to the second circuit and repeat the procedure until you have mapped all the circuits *(inset)*. Include 240-volt circuits that serve major appliances and are controlled by double breakers or fuse blocks. Use the information recorded on the floor plans to label the various breakers or fuses *(left)*. Write a short description of each circuit beside its corresponding number. Leave unused breakers turned off and unused fuse sockets empty and label them clearly.

INSPECTING THE CIRCUITS

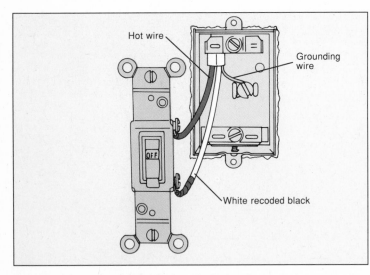

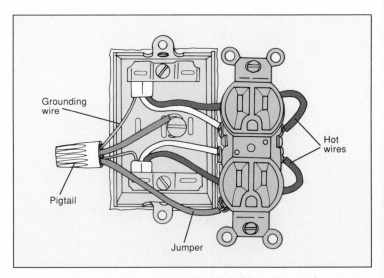

Familiarizing yourself with a typical circuit. Begin your inspection at the service panel. With one hand, open the door to reveal the fuses or circuit breakers. You should have 15-, 20- and 30-ampere fuses or circuit breakers in the panel. Check the condition of the cables entering the panel. If you discover any cracked or frayed sheathing, call for a professional inspection of your house wiring. Note that the wiring behind the panel cover is live and dangerous, to be serviced only by an electrician.

Turn off power to one circuit by removing the fuse or tripping the circuit breaker *(page 18)*. Examine each switch and outlet cover on the circuit. Check for warm or discolored cover plates, indicating poor connections at the device. Next, look inside two typical boxes, following the instructions for switches *(page 70)* and outlets *(page 86)* to free each device from its box and confirm that the power is off. Examine the

condition of the switch or outlet and look for dirty or loose wire connections at the terminals. (If the switch and outlet are stamped CO/ALR and the bare wire ends appear white rather than copper-colored, you probably have aluminum wiring. Call an electrician to inspect your system.) Note the number of cables entering the box. One cable *(above, left)* means that the switch or outlet is located at the end of the circuit; two or more cables *(above, right)* indicate that the device is installed mid-circuit. Examine the color coding of the wires in the box. Locate the black (hot) wire, the white (neutral) wire and the bare grounding wire leading from each cable. When the white wire is used to carry current, it is recoded black with electrical tape or black paint *(above, left)*. Sometimes a short piece of insulated wire, called a jumper, is used to link two or more wires to a screw terminal. It is attached to the other wires in a pigtail connection *(above, right)*.

AN ELECTRICAL FLOOR PLAN

Labelling the service panel. Use these charts to map your home's electrical circuits, following the instructions on page 20. Sketch the floor plan of your house on the grids that follow, then locate the various switches, outlets and fixtures on each circuit. Summarize the information on the label at right and use it as a quick-reference guide before beginning any electrical work.

MAIN SERVICE PANEL			
Circuit		**Circuit**	
1		15	
2		16	
3		17	
4		18	
5		19	
6		20	
7		21	
8		22	
9		23	
10		24	
11		25	
12		26	
13		27	
14		28	

In the basement. Sketch the floor plan of your basement on the grid at right. Mark the location of the main service panel and any sub-panels. Using the symbols at the top of the grid, indicate all outlets, switches and light fixtures. Include 240-volt outlets for major appliances such as the water heater, clothes dryer and central air conditioner. (A 240-volt circuit is protected by a double breaker or fuse block.) Locate junction boxes, often found on basement ceiling joists, and trace the low-voltage wiring of the doorbell system. Note that outlets found in workshops are often on a 120/240-volt split circuit *(page 95)*. If a cable travels outside the house through the foundation wall, include any outdoor outlets, switches or lighting fixtures in the sketch.

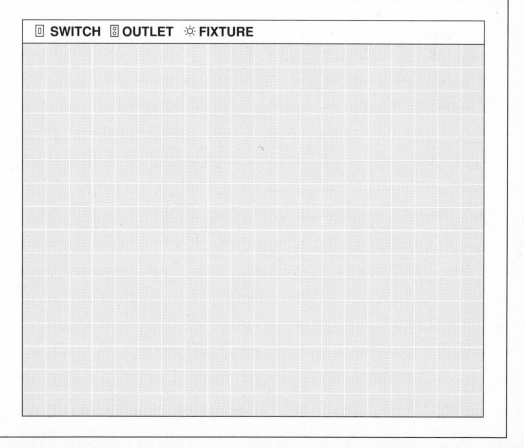

SWITCH   OUTLET   ☼ FIXTURE

AN ELECTRICAL FLOOR PLAN (continued)

On the first floor. Sketch the plan of the first floor on the grid at right. Using the symbols at the top of the grid, indicate all outlets, switches and light fixtures. The kitchen will be served by several circuits, including dedicated circuits for large appliances. Countertop outlets are often on a 120/240-volt split circuit *(page 95)*. Look for outdoor or garage circuits that extend from a circuit on this floor.

⬓ **SWITCH** ⬓ **OUTLET** ☼ **FIXTURE**

On the second floor. Sketch the plan of the second floor on the grid at right. Using the symbols at the top of the grid, indicate all outlets, switches and light fixtures. This area of the house is usually served by several 120-volt, 15-ampere circuits.

⬓ **SWITCH** ⬓ **OUTLET** ☼ **FIXTURE**

CALCULATING ELECTRICAL LOAD

Determining a circuit overload. If you suspect that a circuit is overloaded, it is a simple matter to calculate the existing load, then compare this to the capacity of the circuit. The maximum load is indicated by the amperage rating of the fuse or circuit breaker. For example, a general lighting circuit would have a maximum load of 15 amperes.

To calculate the existing load on a circuit, list all the fixtures and appliances on the circuit and the wattage rating for each device. This information is usually found on a sticker near the socket of a lamp or lighting fixture, or on a small plate on the back or bottom of an appliance *(right)*. Typical wattage ratings are listed in the chart below. Add the wattage ratings for all appliances and fixtures on the circuit, then divide by 120 volts to convert to amperes. If the total is higher than the capacity of the circuit, the circuit is overloaded. Move a heavy-drawing appliance to another circuit or have an electrician run a new circuit from the service panel. When planning a circuit extension or adding a new appliance to the circuit, follow the same procedure to determine the existing and maximum loads of the circuit.

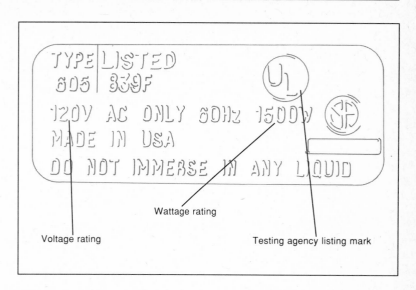

Voltage rating

Wattage rating

Testing agency listing mark

TYPICAL APPLIANCE LOADS

APPLIANCE	APPROXIMATE WATTAGE RATING	APPLIANCE	APPROXIMATE WATTAGE RATING	APPLIANCE	APPROXIMATE WATTAGE RATING
Household appliances and equipment		**Bedroom and bathroom appliances**		**Kitchen appliances**	
Ceiling fan	50	Electric blanket	150 - 500	Blender	200 - 400
Clock	5	Hair dryer	400 - 1500	Can opener	150
Computer	150 - 600	Heating pad	60	Coffee grinder	150
Dehumidifier	575	Shaver	15	Coffeemaker	600 - 750
Humidifier (cold mist)	50 - 150	**240-volt appliances**		Food processor	500 - 1500
Iron	1200	Air conditioner	5000	Frying pan	1000 - 1200
Lamp or fixture	25 - 150	Clothes dryer	5000	Kettle	1200 - 1400
Portable heater	1500	Range (per burner)	5000	Mixer	100 - 225
Projector	350 - 500	(oven)	4500	Toaster	800 - 1200
Radio	10 - 100	Water heater	2500 - 5000	Toaster-oven	1500
Sewing machine	100	**Large 120-volt appliances**		Trash compactor	500 -1000
Stereo system	200 - 500	Dishwasher	1200 - 1500	**Power tools**	
Television	150 - 450	Freezer	300 - 600	Drill	360
Typewriter	45	Garbage disposer	300 - 900	Sander	540
Vacuum cleaner	300 - 600	Microwave oven	650	Saw	600 - 1500
Videocassette recorder	50	Refrigerator	150 - 300	Soldering iron	150

GROUNDING FOR SAFETY

Grounding at the service panel. Grounding is a safety precaution built into every home electrical system. At the service panel, the main grounding wire is connected to a metal water pipe and grounding rod that is buried in the earth, providing excess current with a direct path to the ground.

In the house wiring, a bare copper or green insulated grounding wire provides an alternate path for leaking current, protecting the circuit from damage and the user from shock. In the example at right, the hot wire supplying current to a lighting fixture has become disconnected from its socket terminal. Since the circuit cannot be completed via the neutral wire, the metal box would become electrified and dangerous. But the grounding wire picks up this leaking current and returns it to the service panel, where it will trip the breaker or blow the fuse.

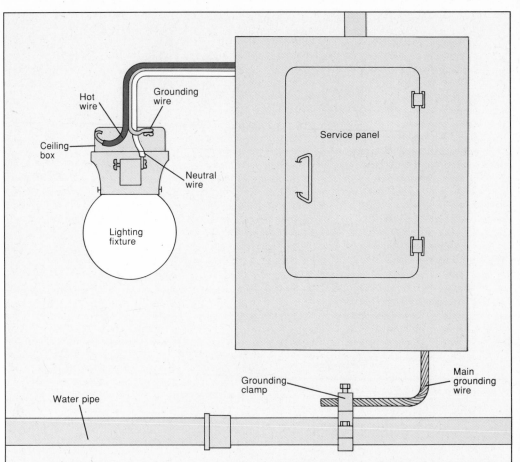

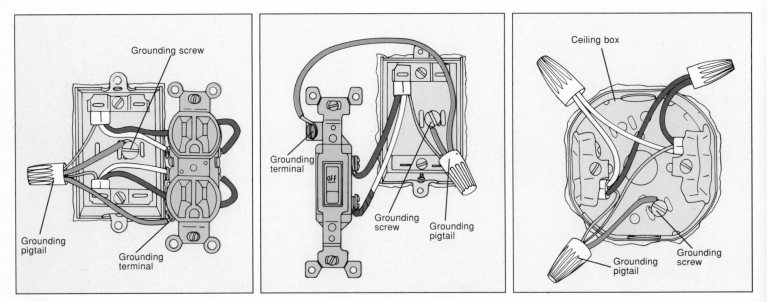

Grounding in outlet, switch and ceiling boxes. The bare copper grounding wire in the cable provides protection against current leakage at an electrical box. It is attached to the grounding screw at the back of a metal box or to the grounding terminal on a switch or outlet, and grounds the box, the mounting strap, the device or fixture and all three-prong appliances that are plugged into a grounded outlet. When there is more than one cable in the electrical box, grounding is accomplished using jumper wires in a pigtail connection *(page 141)*. In Canada, more than one grounding wire may be attached to the grounding screw at the back of the box. For more information on grounding in boxes, refer to switches *(page 73)* and outlets *(page 89)*.

GROUND-FAULT CIRCUIT INTERRUPTERS

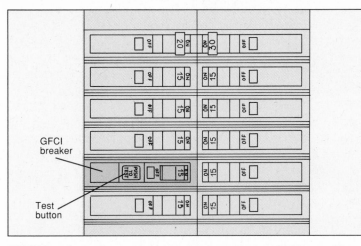

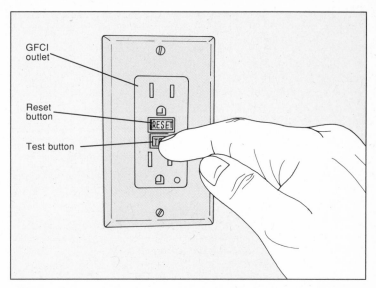

GFCI breakers. Ground-fault circuit interrupters are highly sensitive breakers that measure the current entering and leaving a device along the circuit. If the difference is greater than .005 amperes, the breaker instantly interrupts the flow of current, shutting down the circuit before you can be seriously hurt. A GFCI breaker *(above)* replaces a regular breaker in a service panel. It comes in 15- to 30-ampere versions and is available for both 120- and 240-volt circuits. The GFCI should be checked regularly by pressing the TEST button. If the breaker is good, it will trip; to reset it, flip the toggle back to the ON position.

GFCI outlets. If it is installed at the first outlet box in the circuit, the GFCI outlet *(above)* will protect all outlets along the circuit. The National Electrical Code now requires that new outlets within 6 feet of the kitchen sink and in bathrooms, garages, basements and outdoors must be GFCI-protected. Also available is a portable, plug-in GFCI that fits into any three-slot outlet. To test a GFCI outlet, push the TEST button; the RESET button will pop out. Reactivate the GFCI by pressing the RESET button. To service a GFCI outlet, see page 96.

POLARIZATION IN LAMPS AND FIXTURES

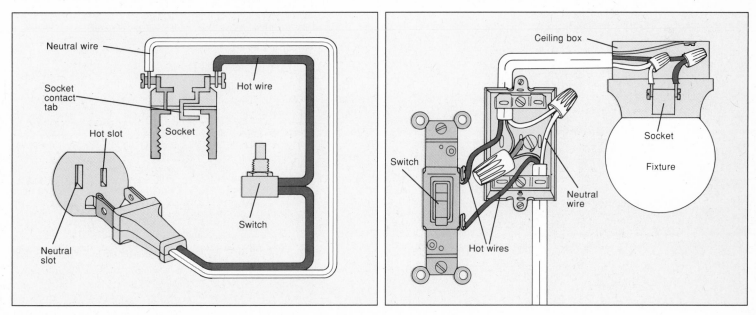

Stopping current at the switch. In a properly wired lamp or lighting fixture, the switch interrupts the hot wire that carries current forward, ensuring that no electricity flows through the lamp or fixture when the switch is turned off. If polarity is reversed, an exposed socket can give a shock even though the switch is off. Polarization in a lamp *(above, left)* begins at the wall outlet. The narrow slot of the outlet is hot; when a polarized lamp cord is plugged into the outlet, power enters the plug through the narrow prong and is transmitted through the hot, unmarked wire to the brass socket terminal. In a lighting fixture *(above, right)*, current flows from the hot wire at the wall switch to the brass terminal of the fixture socket.

FLOOR AND TABLE LAMPS

You may own floor or table lamps, modern or antique, but the parts and wiring are basically the same. Once you know how to fix it, you need never throw out a lamp that does not work.

When a lamp is not working properly, check first for the simplest cause—a loose or burned-out bulb, a blown fuse or tripped circuit breaker, or a plug that has been knocked out of the wall outlet. Confirm that the outlet is not at fault by plugging another lamp into it. Finally, check the lamp socket, switch, plug and wiring.

After the bulb, the socket is the part that fails most often. The socket may have a simple ON/OFF switch, or a three-way switch that offers several lighting levels. Since the wire connections to each type of socket are identical, you may want to consider replacing a one-way socket with a three-way. The switch may also be found in the lamp base or cord. In this case the socket and switch can be repaired separately.

Always unplug the lamp before beginning a repair and, while you're at it, inspect the cord and plug. To avoid damaging the lamp cord, always pull on the plug when removing it from the outlet. Check the plug regularly; even the most carefully handled plug will eventually wear out, and should be replaced as soon as it shows signs of damage.

If you suspect the lamp wiring is faulty, examine the cord for frayed, cracked or bare wires, or for loose connections at the plug. It is common to find a broken or damaged wire at the plug or lamp base. Never splice a worn lamp cord; replace it completely.

If you have never wired anything before, lamps are a good place to start. When you replace the cord, select one the same gauge as the old cord; most lamps use No.18/2 lamp cord (also called zip cord). To determine how much cord to buy, measure the old cord and add the height of the lamp; then add at least

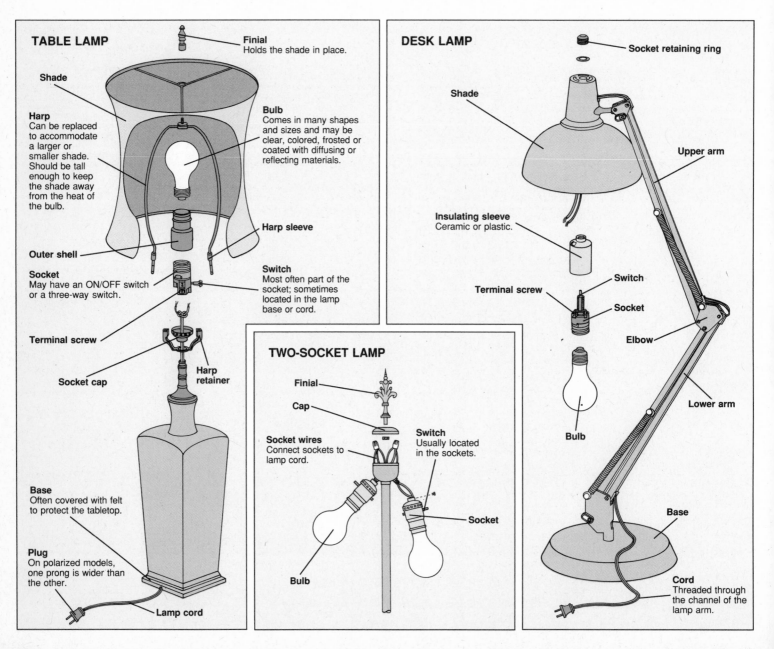

TABLE LAMP

Finial
Holds the shade in place.

Shade

Harp
Can be replaced to accommodate a larger or smaller shade. Should be tall enough to keep the shade away from the heat of the bulb.

Bulb
Comes in many shapes and sizes and may be clear, colored, frosted or coated with diffusing or reflecting materials.

Outer shell

Harp sleeve

Socket
May have an ON/OFF switch or a three-way switch.

Switch
Most often part of the socket; sometimes located in the lamp base or cord.

Terminal screw

Socket cap

Harp retainer

Base
Often covered with felt to protect the tabletop.

Plug
On polarized models, one prong is wider than the other.

Lamp cord

TWO-SOCKET LAMP

Finial

Cap

Socket wires
Connect sockets to lamp cord.

Switch
Usually located in the sockets.

Socket

Bulb

DESK LAMP

Socket retaining ring

Shade

Upper arm

Insulating sleeve
Ceramic or plastic.

Switch

Terminal screw

Socket

Elbow

Lower arm

Bulb

Base

Cord
Threaded through the channel of the lamp arm.

another foot. Replace the entire cord and, for maximum safety, replace the plug at the same time. (Sometimes you can buy the cord and plug as one piece).

Lamp cord has many fine wire strands beneath its plastic insulation. When making connections at terminal screws, or to other wires, twist the strands together to be sure that there are no stray wire ends *(page 139)*.

When rewiring a lamp, it is also best to polarize it *(page 25)* so that the lamp switch interrupts the current-carrying hot wire. That way, when the lamp is turned off, there will be no power flowing through it. You will need a polarized plug, which has one prong wider than the other, and polarized cord, which is ribbed or color-coded on one side or has a thread

evident in the strands of one wire. This is the neutral side, and should be connected to the silver terminal of the socket and the wide prong of the polarized plug.

One of the newest developments in home lighting is the halogen light. Such lamps use a high-pressure, high-temperature bulb that supplies the purest lighting possible and is more efficient than the standard incandescent bulb. You can often light an entire room effectively with one halogen lamp. Because the design and parts of halogen lamps are not standardized, however, it is best to take these lamps to a dealer for repairs for all but the simplest problems. The diagram of a typical halogen lamp and steps for pinpointing a problem appear on page 42.

TROUBLESHOOTING GUIDE

SYMPTOM	POSSIBLE CAUSE	PROCEDURE
Bulb flickers or does not light	Plug loose or unplugged	Check plug in wall outlet
	Bulb loose	Tighten bulb
	Bulb burned out	Replace bulb
	Wall outlet faulty	Test lamp in another outlet. If it lights, service outlet *(p. 86)*
	Socket contact tab bent	Pry up contact tab *(p. 30)* □○
	Socket contact tab damaged	Replace socket *(p. 30)* □○
	Socket faulty	Replace socket *(p. 30)* □○
	Lamp switch faulty	Test and replace switch *(socket with switch, p. 30* □○*; switch in lamp base, p. 33* □○*; flat-cord switch, p. 34* ◨○*; round-cord switch, p. 34* ◨○*)*
	Plug faulty	Replace plug *(flat-cord plug, p. 36* □○*; quick-connect plug, p. 36* □○*; round-cord plug, p. 37* □○*)*
	Lamp cord faulty	Test and rewire lamp *(one-socket lamp, p. 38* ◨●*; two-socket lamp, p. 39* ◨●*; desk lamp, p. 41* ◨●*)*
Lamp blows fuse or trips circuit breaker	Overload at service panel	Check load on circuit *(p. 23)*
	Short circuit in lamp cord	Test and rewire lamp *(one-socket lamp, p. 38* ◨●*; two-socket lamp, p. 39* ◨●*; desk lamp, p. 41* ◨●*)*
	Plug faulty	Replace plug *(flat-cord plug, p. 36* □○*; quick-connect plug, p. 36* □○*; round-cord plug, p. 37* □○*)*
	Socket faulty	Test and replace socket *(p. 30)* □○
Bulb burns out too quickly	Heavy use of lamp	Install long-life bulb or bulb-life extender *(p. 28)*
Shock when changing bulb	Socket faulty	Replace socket *(p. 30)* □○
Shock when plugging or unplugging lamp	Plug faulty	Replace plug *(flat-cord plug, p. 36* □○*; quick-connect plug, p. 36* □○*; round-cord plug, p. 37* □○*)*
	Lamp cord faulty	Rewire lamp *(one-socket lamp, p. 38* ◨●*; two-socket lamp, p. 39* ◨●*; desk lamp, p. 41* ◨●*)*
HALOGEN LAMPS		
Lamp does not light	Bulb burned out	Replace bulb *(p. 42)* □○
	Lamp cord faulty	Inspect wire at plug and lamp base. Take for repairs
	Transformer, socket or fuse faulty	Test to locate problem *(p. 42)* ◨○▲, then take lamp for repairs

DEGREE OF DIFFICULTY: □ **Easy** ◨ **Moderate** ■ **Complex**
ESTIMATED TIME: ○ **Less than 1 hour** ◖ **1 to 3 hours** ● **Over 3 hours**   ▲ **Multitester required**

INCANDESCENT BULBS AND SOCKETS

The right bulb for the job. Most household lamps and lighting fixtures operate with incandescent bulbs. When the switch is turned on, current travels through the wiring to the socket, then to the tungsten filament inside the bulb. Light is produced when the filament is heated and glows. Fluorescent lamps and lighting fixtures work on a different principle, as described on page 60.

Indoor incandescent lamps and fixtures use a standard A-type bulb with a threaded base that screws into a medium-base socket. The three-way bulb, similar in appearance to the A-type bulb, requires a special socket. Two filaments inside this type of bulb light up, one at a time for low- and medium-intensity lighting, or together for maximum lighting. Track and recessed fixtures usually require reflector bulbs, which provide a bright, concentrated beam.

For the demands of outdoor lighting *(page 120)*, there are special bulbs made of shatterproof glass designed to withstand sudden temperature changes. Most outdoor lamps and fixtures come with medium-base, insulated sockets *(page 29)*, and take A-type bulbs. Other types of outdoor lighting include Christmas lights, which fit into small, threaded candelabra sockets, and low-voltage bulbs that fit screw-type or bayonet sockets.

Some light bulbs are sold in clear, frosted and coated versions, providing different qualities of light. Bulbs are also available in different wattages; the higher the wattage of the bulb, the more light it will provide but the more electricity it will use. For stairways, hallways, outdoor fixtures and other hard-to-reach areas, consider buying long-life bulbs; the filament of these bulbs burns at a lower temperature, increasing the life of the bulb up to three times that of a regular bulb. Another option is the bulb-life extender, a small button-like disc that can add years to the life of a standard A-type bulb by limiting the surge of electricity to the bulb. To install a bulb-life extender, simply peel off the backing and stick it to the base of the bulb before threading it into the socket.

When buying a replacement bulb, consider your lighting requirements, then check the size of the socket *(page 29)* and its maximum recommended wattage. The wattage information is usually found on a sticker near the socket of most lamps and fixtures. Never use a bulb that exceeds the socket's maximum recommended wattage. If you do, the heat produced could melt a plastic globe or shade, or damage the wire insulation. Because air space in a globe is essential, avoid using a bulb that is too large for the lamp or fixture.

BULB TYPE		APPLICATION
A-type		Most common type of light bulb. Used in standard lamps and fixtures. Weather-resistant outdoor version also available. For medium-base socket; 4-300 watts.
Three-way		For lamps with three-way sockets; provides low, medium and high light intensities. For medium-base socket; 30/70/100 and 100/200/300 watts.
Long-life		Where a longer-lasting bulb is required. For medium-base socket; 40-150 watts.
T (tubular)		Found in medium-intensity desk lamps, or used as a showcase bulb, mounted in a fixture above a picture frame or in the canopy of an aquarium. For medium-base socket; 15-150 watts.
Candelabra		For lamps and fixtures where low light or accent lighting is desired. For medium-base and candelabra socket; 15-60 watts.
G (globe)		A decorative bulb that does not require a shade. Used around makeup mirrors and in hanging fixtures. For medium-base and candelabra socket; 40-150 watts.
Night light		Small, low-wattage bulb used in plug-in fixtures to illuminate hallways at night. Also used as Christmas lights. For medium-base and candelabra socket: 4, 7, 7 1/2 watts.
R (reflector)		Also called spotlight. A built-in reflector directs the beam. Used in adjustable fixtures (such as track lights) where directional lighting is required. For medium-base socket; 15-60 watts.
PAR (parabolic aluminized reflector)		Also called floodlight. Bright, high-wattage bulb used in track lighting, recessed fixtures and outdoors. For medium-base socket; 25-250 watts.
ER (ellipsoidal reflector)		Used in recessed, downlight fixtures. For medium-base socket; 50-120 watts.
Low-voltage		Used in indoor fixtures or temporary outdoor fixtures. Requires a transformer to reduce 120-volt household current. For medium-base candelabra and bayonet socket; 6-16 volts.

INCANDESCENT BULBS AND SOCKETS

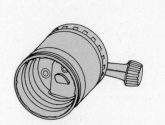
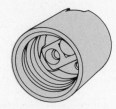
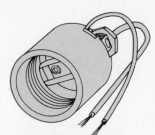

Variations in sockets. Most household lamps and fixtures are fitted with a medium-base socket designed for use with bulbs up to 300 watts. Sockets with smaller bases, such as the intermediate base and the tiny candelabra socket, are designed to hold decorative, low-wattage bulbs. Low-voltage systems also use smaller sockets. The examples at left represent some of the most common variations.

Medium-base lamp socket
Normally made of metal, but can also be plastic; may contain a simple ON/OFF switch or a three-way switch; found in floor and table lamps and some chandeliers.

Porcelain fixture socket
Attached to the fixture with an inside mounting screw. This model has one brass and one silver terminal screw.

Plastic fixture socket
Attached to the fixture with an external mounting strap. This model has preattached wires.

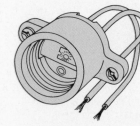

Sockets come with two screw terminals or two preattached wires that are soldered or riveted on. The outer casing is made of metal, plastic or porcelain. For outdoor use, plastic, porcelain and rubber offer durability, resistance to dampness and maximum safety. All sockets have a metal tube, often threaded, and a contact tab inside the socket.

Two-part socket
Separates into two parts that are fitted into a socket hole from each side and screwed together; provides easy installation.

Low-voltage socket
Small-based socket designed to receive low-voltage bulbs. This model has a bayonet base with preattached wires; may also have a threaded metal base.

Outdoor socket
Exterior-grade porcelain socket; may also be plastic or rubber. This model has two external mounting tabs.

Sockets also vary in the way they are mounted to the fixture. A lamp socket, for instance, has a cap that fits onto the center pipe of a lamp. Lamp sockets are also found on chandeliers, where they may have pull-chain, twist or push-in switches.

Fixture sockets may be secured by an inside mounting screw or an external mounting strap. Low-voltage sockets may simply be wedged into the fixture. A two-part socket fits into the fixture from the top and bottom and screws together.

REMOVING A BROKEN BULB

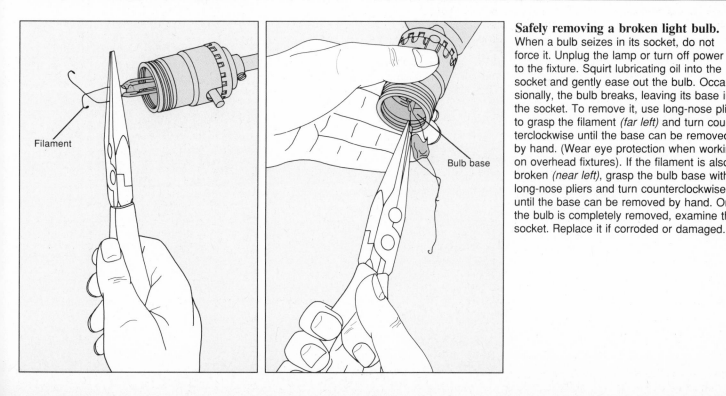

Filament

Bulb base

Safely removing a broken light bulb.
When a bulb seizes in its socket, do not force it. Unplug the lamp or turn off power to the fixture. Squirt lubricating oil into the socket and gently ease out the bulb. Occasionally, the bulb breaks, leaving its base in the socket. To remove it, use long-nose pliers to grasp the filament (far left) and turn counterclockwise until the base can be removed by hand. (Wear eye protection when working on overhead fixtures). If the filament is also broken (near left), grasp the bulb base with long-nose pliers and turn counterclockwise until the base can be removed by hand. Once the bulb is completely removed, examine the socket. Replace it if corroded or damaged.

TESTING AND REPLACING THE SOCKET

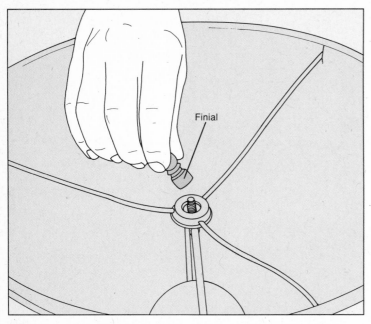

1 **Removing the shade.** Unplug the lamp. If there is a finial securing the shade to the harp, unscrew it *(above)*, then lift off the
shade and unscrew the light bulb. Screw the finial back onto the
harp to avoid losing it.

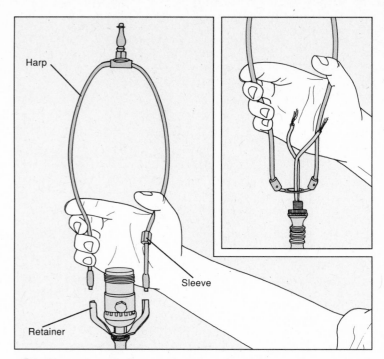

2 **Removing the harp.** Most one-socket lamps have a harp that
is held to a retainer by metal sleeves. To remove it, push up the
sleeves, then squeeze the harp and lift it out of the retainer, as
shown. On some lamps, the harp cannot be lifted off the lamp until the
socket is removed *(inset)*.

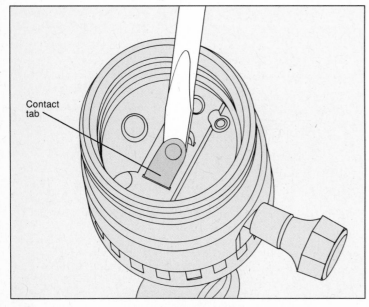

3 **Cleaning and adjusting the contact tab.** With a flat-tipped
screwdriver, scratch any dirt off the surface of the contact tab. If
the tab is corroded or broken, remove and replace the socket
(next step). If you suspect the tab does not protrude enough to make
contact with the bottom of the bulb, use a screwdriver to pry the end
up slightly, as shown, and try the bulb again. The contact tab is deli-
cate; prying it up more than a few times may break it. If the bulb still
does not light, or if the contact tab snaps, unplug the lamp again,
unscrew the bulb and go to step 4.

4 **Removing the outer shell and insulating sleeve.** Look for the
word "PRESS" on the outer shell of the socket near the switch,
then press hard with your thumb, as shown. You may also have
to wiggle the cap slightly to release its grip; do not twist it. Lift the shell
free. Slip off the cardboard insulating sleeve if it does not lift off with
the shell. If the sleeve is damaged, the socket must be replaced.

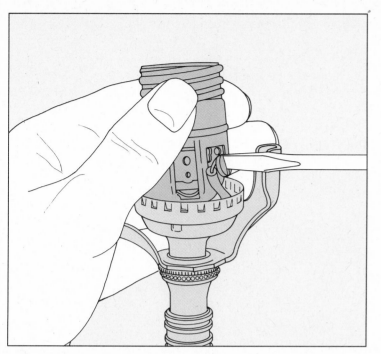

5 **Detaching the socket.** On lamps that have no terminal screws, clip the wires near the socket, lift the socket off and go to the next step. If there are screw terminals, check for a loose wire connection; if a wire is loose, rehook it around the terminal screw and tighten the screw. Then screw in the light bulb, plug in the lamp and turn it on to test. If the lamp still does not light, unplug it and remove the bulb. Detach the socket wires, as shown, and lift the socket up and off the lamp.

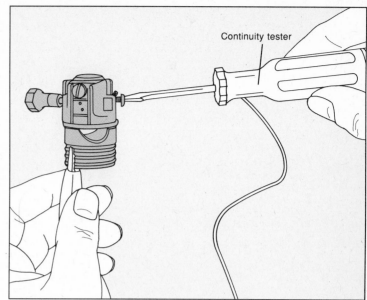

Continuity tester

6 **Testing the socket.** If the socket has screw terminals, place the alligator clip of a continuity tester on the threaded metal base and touch the tester probe to the silver (neutral) terminal, as shown. The tester should light. If the socket tests OK, check the switch in the socket *(page 32)*, or reinstall the socket and check the switch in the lamp base *(page 33)* or cord *(page 34)*. If the tester does not light, go to the next step.

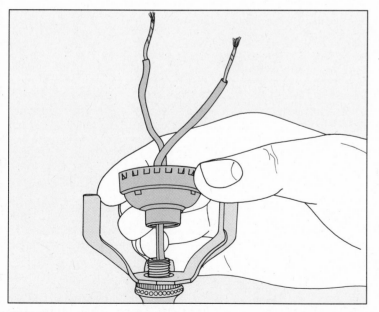

7 **Removing the socket cap.** To remove the socket cap from the lamp, first untie the cord if it is knotted. Loosen the setscrew at the base of the cap, if there is one, then unscrew the cap. Pull it off the center pipe, as shown. If the socket cap holds the lamp body to the center pipe, support the lamp so that the lamp does not come apart. While the lamp is disassembled, inspect its cord. If the cord is in poor condition, rewire the lamp *(pages 38, 39 and 41)*.

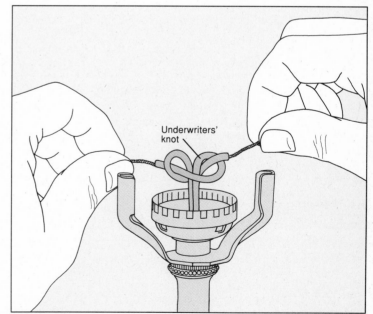

Underwriters' knot

8 **Replacing the socket cap.** Buy a replacement socket with screw terminals. New sockets fit most lamps and come with a socket cap. Thread the lamp cord through the cap, then screw the cap onto the center pipe. If the cord is new, part it about 2 inches back and strip the insulation from the separated ends *(page 138)*. To protect the terminals from strain, tie an Underwriters' knot with the two wire ends, as shown.

TESTING AND REPLACING THE SOCKET (continued)

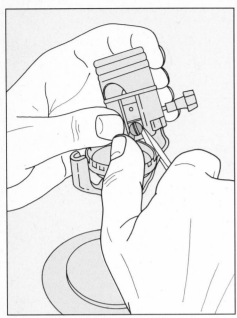

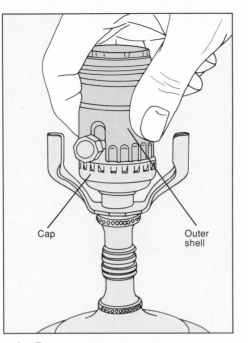

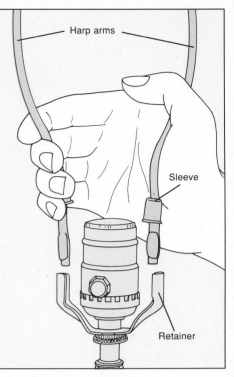

9 **Wiring the new socket.** Twist the strands of each exposed wire end clockwise so that there are no frayed ends. Then, using a screwdriver, hook each wire end around a terminal screw, as shown. Make sure the neutral wire of the cord (the ridged or marked wire) connects to the silver socket terminal and the wide plug prong. The hot or unmarked wire of the cord connects to the brass socket terminal and the narrow plug prong. Tighten the screws, making sure there are no stray wire strands.

10 **Installing the socket.** Slip the insulating sleeve and outer shell onto the socket, fitting the notched opening over the socket switch, as shown. Push the shell into the rim of the cap until it snaps into place. Screw in the bulb, then plug in the lamp and turn it on. If it works, reinstall the harp and lamp shade *(next step)*. If it doesn't, remove the bulb and test the switch *(below, pages 33 and 34)*.

11 **Reassembling the lamp.** Squeeze the arms of the harp together and fit them back into the retainer, as shown, then lower the sleeves in place. Screw in the light bulb and reattach the shade and finial.

TESTING A SOCKET SWITCH

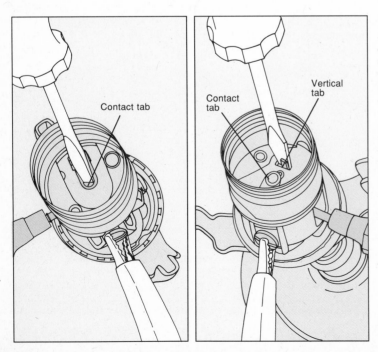

Testing for continuity. If a socket switch seems loose when turned on and off, or if the bulb flickers as the lamp is jiggled, the switch should be checked. Unplug the lamp, remove the shade, unscrew the bulb and lift off the harp, if any. Lift out the socket shell and insulating sleeve *(page 30)* to expose the terminals on the side of the socket. To test a regular ON/OFF switch *(far left)*, place the alligator clip of a continuity tester on the brass terminal and touch the tester probe to the socket contact tab. The tester should light when the switch is on, but not when the switch is off. To test the switch on a three-way socket, place the alligator clip of the continuity tester on the brass screw terminal and touch the tester probe to the small vertical tab in the base of the socket, *(near left)*, then to the contact tab. Turn the switch to the three ON positions, then to the OFF position. The tester should light as follows:

POSITION:	The tester should light when the probe touches:
FIRST	Vertical tab
SECOND	Contact tab
THIRD	Both tabs
OFF	Neither tab

If the switch is faulty, detach and replace the socket *(page 30)*. If the switch tests OK and the socket is good, test the cord and plug *(pages 38, 39 and 41)*.

REPAIRING A SWITCH IN THE LAMP BASE

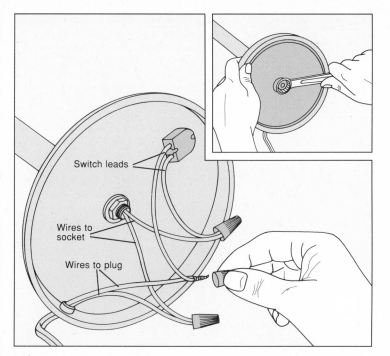

Switch leads

Wires to socket

Wires to plug

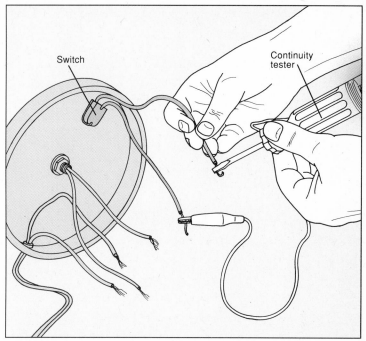

Switch

Continuity tester

1 Access to the switch leads. Flip the switch while the lamp is plugged in. If the switch feels loose when it is turned on and off, or the bulb flickers as the switch is jiggled, the switch is probably faulty. Unplug the lamp and set it on its side. Carefully peel back the protective cover, which is usually felt. If there is a bottom plate, use a wrench to remove the locknut holding it in place *(inset)*, then pull off the plate. If the switch is connected to the lamp cord with wire caps, unscrew the caps, as shown, then untwist the wires.

2 Testing the switch. Locate the two switch leads. Place the alligator clip of a continuity tester on the bare end of one lead and touch the tester probe to the end of the other lead *(above)*. The tester should light when the switch is in one position, but not the other. If the switch fails the test, replace it.

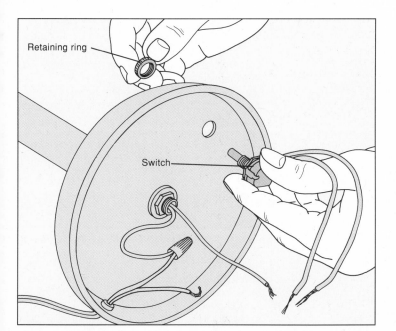

Retaining ring

Switch

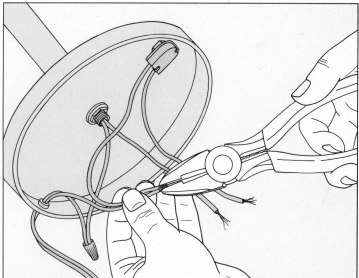

3 Removing the switch. Unscrew the retaining ring on the lamp base, then pull the switch out through the bottom of the base, as shown. Take the switch to a hardware store or electrical supplier and buy a compatible replacement.

4 Installing the new switch. Set the new switch into the lamp base from the bottom and tighten the retaining ring to hold it in place. Connect the unmarked wire from the plug to the unmarked switch lead *(above)*. Next, connect the marked wire from the plug to the marked socket wire. Then connect the marked switch lead to the unmarked socket wire. Screw a wire cap onto each connection. Fit the bottom plate back on and tighten the locknut. Secure the protective cover to the lamp base with cloth glue.

REPAIRING A FLAT-CORD SWITCH

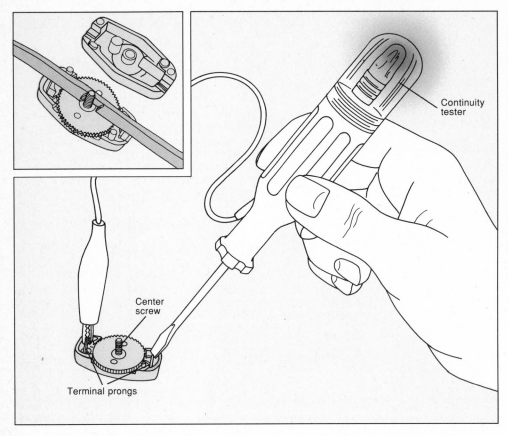

Continuity tester

Center screw

Terminal prongs

Testing and replacing the switch. Unplug the lamp. Unscrew the switch cover and remove the switch from the lamp cord. Place the alligator clip of a continuity tester on one terminal prong and touch the tester probe to the other prong *(left)*. The tester should light only when the switch is in the ON position. If the switch is defective, buy a new flat cord and flat-cord switch. Rewire the lamp *(pages 38, 39 and 41)*. Then choose the most convenient place along the cord to install the new switch. Unscrew the switch cover and separate the switch in two pieces. Using a utility knife, cut a notch large enough to accommodate the center screw of the switch. Set the cord into the switch *(inset)*, then place the switch cover on the cord, making sure that both prongs of the switch pierce the same wire of the cord. If the lamp is polarized, the prongs must pierce the hot (unmarked) wire of the lamp cord. Screw on the switch cover securely.

REPAIRING A ROUND-CORD SWITCH

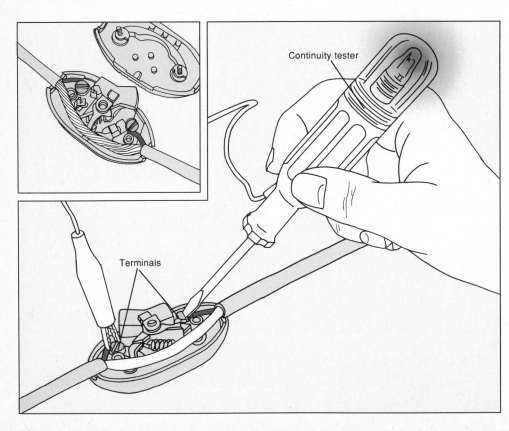

Continuity tester

Terminals

Testing and replacing the switch. Unplug the lamp. Unscrew the switch cover to reveal the terminal connections *(inset)*. Loosen the terminal screws to release the wires and remove the switch from the lamp cord. Place the alligator clip of a continuity tester on one of the switch terminals and touch the tester probe to the other terminal *(left)*. The tester should light when the switch is in one position but not in the other position. If the switch fails this test, buy a new round cord and switch. Rewire the lamp *(pages 38, 39 and 41)*, then choose the most convenient place along the cord to install the new switch. Unscrew the switch cover. Using a utility knife, part the cord, then strip the cord sheathing to expose a section of neutral (marked) wire large enough to accommodate the new switch. Clip the hot (unmarked) wire and strip back the insulation on both ends. Attach each hot wire to a terminal screw. Screw on the switch cover securely.

LAMP DIMMER SWITCHES

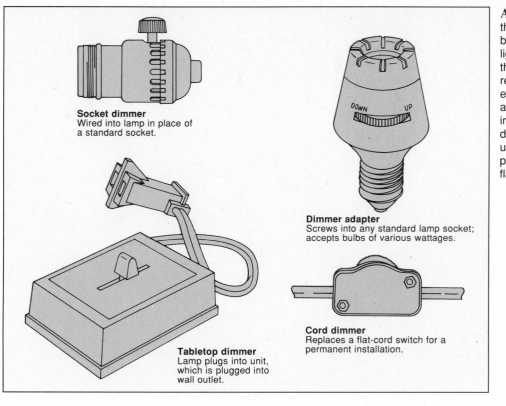

Socket dimmer
Wired into lamp in place of a standard socket.

Dimmer adapter
Screws into any standard lamp socket; accepts bulbs of various wattages.

Cord dimmer
Replaces a flat-cord switch for a permanent installation.

Tabletop dimmer
Lamp plugs into unit, which is plugged into wall outlet.

Adding a dimmer switch. Installing one of these devices to your lamp can provide flexibility and create dramatic effects in home lighting; it can also save energy and extend the life of your bulb. Use a dimmer socket to replace a regular lamp socket, or modify the existing socket by screwing in a dimmer adapter. To use a tabletop dimmer, plug it into an outlet, then plug the lamp into the dimmer plug. For a more permanent solution, use a cord dimmer that separates into two parts and is attached in the same way as a flat-cord switch *(page 34).*

PLUGS

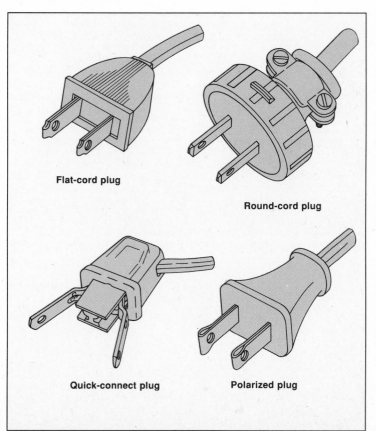

Flat-cord plug

Round-cord plug

Quick-connect plug

Polarized plug

Servicing 120-volt plugs. A glance will tell you whether a plug's casing is cracked or the prongs are loose, bent or corroded. In any of these cases, the plug should be replaced. If a test of the plug and cord *(pages 38, 39 and 41)* shows no continuity, tighten connections between the two and test again. If there is still no continuity, rewire the lamp and replace the plug *(page 36).*

As a rule of thumb, replace a defective plug with one of the same type; also select the same type of cord (most lamps use No. 18/2 lamp wire, called zip cord). The most common type of plug is the flat-cord plug, connected to the lamp cord with terminal screws. An Underwriters' knot is often made in the cord to guard the terminals from strain.

Some lighting devices (such as swag lamps that hang by their cords) need the extra strength of a round cord, which must be fitted with a special plug. If the insulating disc is missing from a round-cord plug, the plug is unsafe and should be replaced.

By far the easiest plug to install is the quick-connect plug. However, its connections are not as sturdy as the other types, and it should be used only for lamps that are seldom unplugged.

When buying a replacement, look for a polarized plug, with one narrow prong and one wide prong. The wider (neutral) prong will not fit into the narrow (hot) slot of modern outlets. By connecting the marked wire of the cord to the wide prong at the plug and the silver terminal at the socket, you can ensure that the lamp switch interrupts the hot wire, and no current flows to the light when the switch is off *(page 25).*

REPLACING A FLAT-CORD PLUG

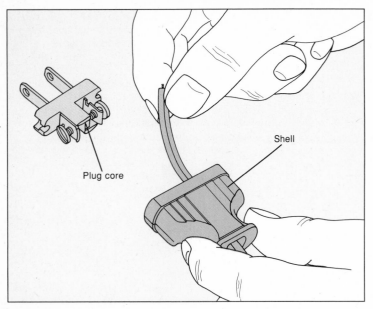

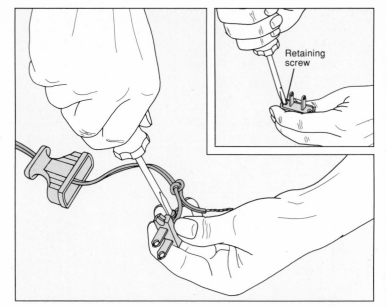

1 **Setting the new plug in place.** Unplug the lamp. Snip the faulty plug off the old wire with a utility knife or diagonal-cutting pliers. For best results, remove the lamp cord and rewire the lamp *(pages 38, 39 and 41)*. If the new plug has a core and shell, separate them and slip the cord through the shell *(above)*. Peel apart the two wires at the end of the cord and strip about 3/4 inch of insulation from each wire end *(page 138)*. To protect the terminals from strain, tie an Underwriters' knot with the two wire ends.

2 **Connecting the terminals.** Twist the wire strands clockwise and use a screwdriver to hook each wire end around a terminal screw. Attach the unmarked (hot) wire to the side with the narrow prong, and the marked (neutral) wire to the side with the wide prong. Tighten the terminal screws, as shown, making sure there are no stray ends. Close the plug by snapping the core into the shell and tightening any retaining screws *(inset)*.

INSTALLING A QUICK-CONNECT PLUG

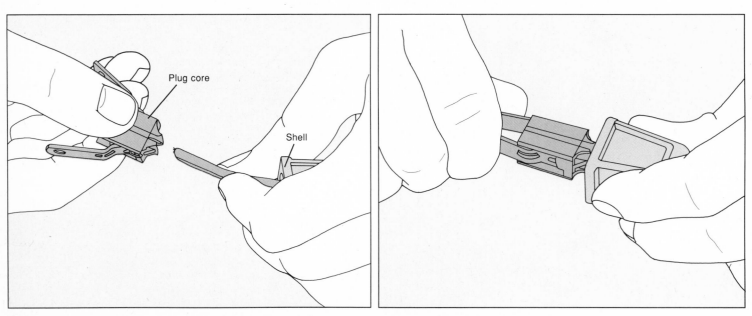

Taking a wiring shortcut. Unplug the lamp. Snip the faulty plug off the old wire with a utility knife or diagonal-cutting pliers. For best results, remove the old wire and plug and rewire the lamp *(pages 38, 39 and 41)*. Spread open the prongs of the new plug by hand or, on some quick-connect plugs, by lifting a lever on top of the plug, then insert the cord into the plug core *(above, left)*. Insert the unmarked wire of the cord on the side of the narrow prong. Then squeeze the prongs together by hand or with the lever, piercing the cord. Slide the shell, if any, over the plug *(above, right)*.

SERVICING A ROUND-CORD PLUG

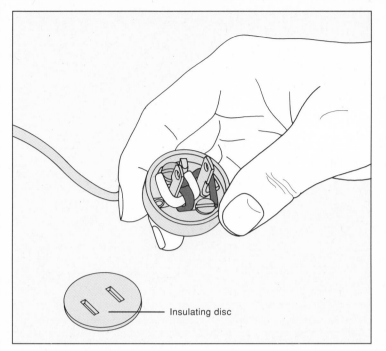

Insulating disc

1 **Removing the old plug.** Unplug the lamp and inspect the plug. If the insulating disc is missing, or the prongs are bent or broken, snip the faulty plug off the old wire with a utility knife or diagonal-cutting pliers. For best results, remove the old wire and rewire the lamp (pages 38, 39 and 41). Then replace the plug (next step).

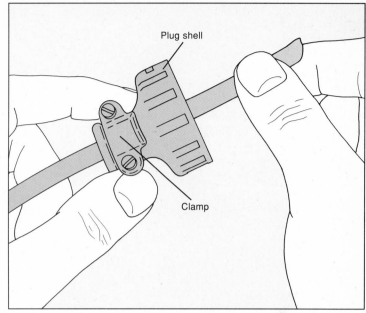

Plug shell

Clamp

2 **Preparing the new plug.** Buy a replacement round-cord plug and pry off the shell with a flat-tipped screwdriver. Slide the shell onto the cord, as shown, and strip 1 1/2 inches of insulation from each wire end (page 138). Tie an Underwriters' knot with the two wire ends.

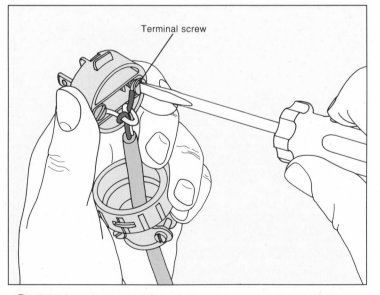

Terminal screw

3 **Making the connections.** Connect the wire ends to the terminal screws of the new plug (above), hooking the white wire around the silver screw and the black wire around the brass screw. Tighten the connections, making sure there are no stray ends.

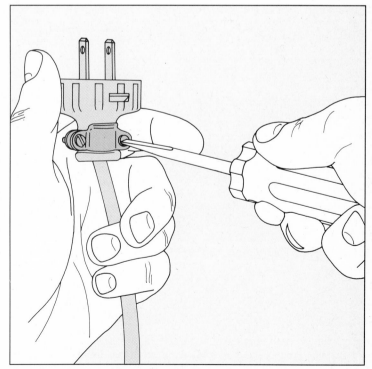

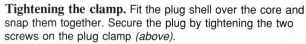

4 **Tightening the clamp.** Fit the plug shell over the core and snap them together. Secure the plug by tightening the two screws on the plug clamp (above).

REWIRING A ONE-SOCKET LAMP

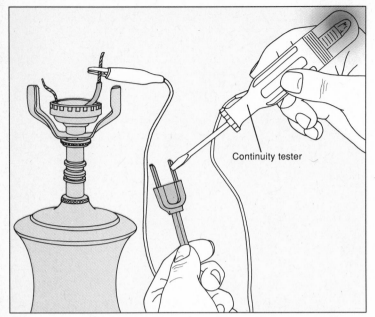

1 **Testing the cord and plug.** Unplug the lamp. Remove the shade, unscrew the bulb and lift off the harp, then remove the socket *(page 30)*. If the plug is not polarized, place the alligator clip of a continuity tester on one wire end and touch the tester probe first to one plug prong *(above)*, then to the other. The tester should light for only one prong. Repeat the test for the other wire end. If the plug is polarized (one prong is wider than the other), place the alligator clip of the continuity tester on the end of the unmarked wire and touch the tester probe to the narrow plug prong. Repeat the test for the wide prong and marked wire. On a good cord and plug, the tester should light for both tests. If the tester does not light, tighten the plug connections and test again. If the cord and plug fail the second test, replace them *(next step)*.

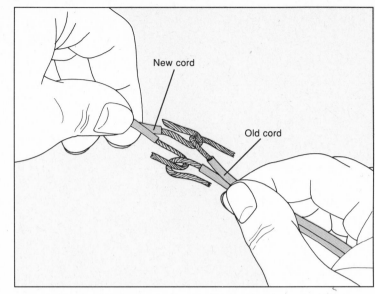

2 **Splicing the old and new cords.** If your lamp has a short, straight pipe, simply remove the old cord and feed in the new one, then go to step 4. If the lamp is tall or you suspect that the center pipe bends inside the lamp, use the old cord to pull the new cord through the lamp. Part the new cord 2 inches back, then strip 3/4 inch of insulation from each wire end *(page 138)*. Twist the strands clockwise, then hook the ends of the old and new cords together *(above)*. Secure the splice by wrapping it tightly with electrical tape so that it will fit through the lamp. If the splice is too thick, unwrap it and use a single wire from each cord *(page 41, step 2)*.

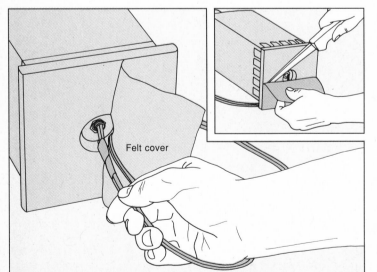

3 **Fishing the new cord.** Using a knife, carefully pry the felt cover away from the lamp base *(inset)*. Peel it back far enough to reveal the cord in the lamp base. (Try to keep the felt in one piece for later regluing.) Feed the splice into the center pipe at the top of the lamp. Then, from the base, pull on the old cord until the splice appears *(above)*. Remove the tape, undo the splice and discard the old cord. Continue pulling the new cord through, leaving enough at the top to attach the socket.

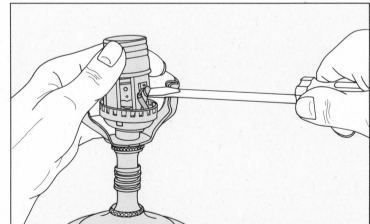

4 **Reassembling the lamp.** Thread the new lamp cord through the socket cap, then screw the cap to the lamp. Part the cord about 2 inches back and strip 3/4 inch of insulation from the separated ends *(page 138)*. Tie an Underwriters' knot with the two wire ends. Hook the unmarked wire around the brass socket terminal and the marked wire around the silver socket terminal, then tighten the connections *(above)*. Slip the insulating sleeve and outer shell onto the socket, and push the shell into the cap. Install a new plug *(page 36)*. Reattach the bulb, harp and shade. Plug in the lamp and test it.

REWIRING A TWO-SOCKET LAMP

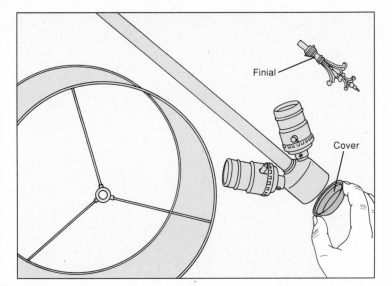

Finial

Cover

1 **Access to the wiring.** Unplug the lamp. Remove the light bulbs, then unscrew the finial and lift off the lamp shade. Unscrew the cover at the junction of the sockets, as shown. Then press on the outer shell of one socket and pull it off with its insulating sleeve. Loosen the terminals on the socket, disconnect the wires and set the socket aside.

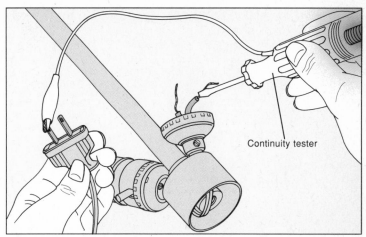

Continuity tester

2 **Testing the cord and plug.** If the plug is not polarized, place the alligator clip of the continuity tester on one plug prong and touch the tester probe first to one wire end *(above)*, then to the other. The tester should light when only one wire end is touched. Repeat the test for the other prong. If the plug is polarized (one prong is wider than the other), place the alligator clip of the continuity tester on the narrow plug prong and touch the tester probe to the unmarked wire at the socket end of the cord. Repeat the test for the wide prong and marked wire. The tester should light in both cases. If it doesn't, tighten the connections and retest. If the cord and plug fail the second test, rewire the lamp *(next step)*. Otherwise, disassemble and test the second socket.

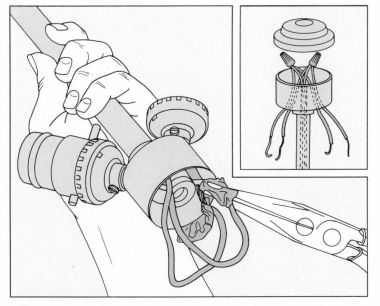

3 **Disconnecting the wires.** Using long-nose pliers, pull out the wires that connect the sockets to the lamp cord *(above)*. Note how the lamp is wired. In most cases, the main lamp cord is parted and each wire is joined with a wire cap to a wire from each socket *(inset)*. The marked wires are joined with one wire cap and the unmarked wires with another. Remove any electrical tape, twist off the wire caps and disconnect the wires.

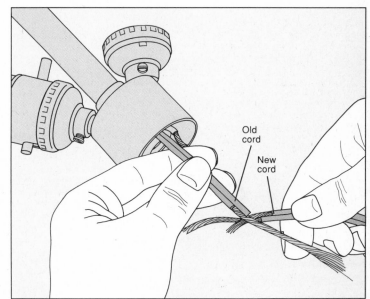

Old cord

New cord

4 **Splicing the old and new cords.** Pull the socket wires out of the lamp and keep them aside to measure new wires in step 6. If your lamp has a short, straight center pipe, simply remove the old cord and feed in the new cord, then go to step 6. If the lamp has a long center pipe, or you suspect that the center pipe bends inside the lamp, use the old cord to pull the new cord through the lamp. Part the new cord about 2 inches back and strip 3/4 inch of insulation from each wire end *(page 138)*. Twist the strands clockwise, then hook the ends of the old and new cords together. Secure the splice by wrapping it with electrical tape, keeping the splice as thin as possible so that it will fit through the lamp. If the splice is too thick, undo it and use a single wire from each cord *(page 41, step 2)*.

REWIRING A TWO-SOCKET LAMP (continued)

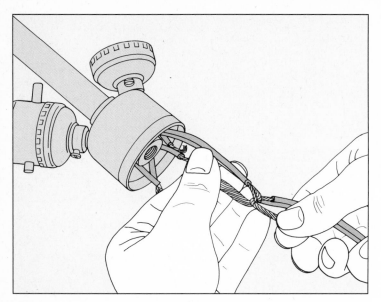

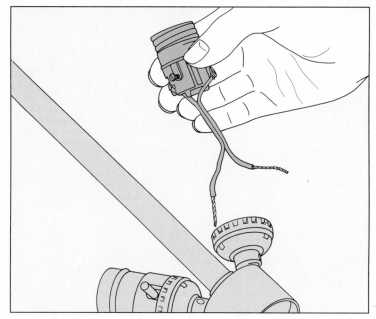

5 **Fishing the new cord.** Feed the splice into the top of the lamp, pushing it into the center pipe. Then pull gently on the old cord at the lamp base until the new cord appears. Disconnect the splice *(above)* and discard the old cord. Install a new plug *(page 36)*.

6 **Replacing the socket.** Cut a piece of lamp cord the same length as the old socket wires, part the cord 1 1/2 inches back and strip both ends *(page 138)*. Wrap the unmarked wire around the brass socket terminal and the marked wire around the silver socket terminal, then tighten the connections. Return the socket to the lamp, as shown, feeding the wires through the socket cap and the top of the lamp. Slip the insulating sleeve and socket shell onto the socket and push the shell into the rim of the cap until it snaps into place.

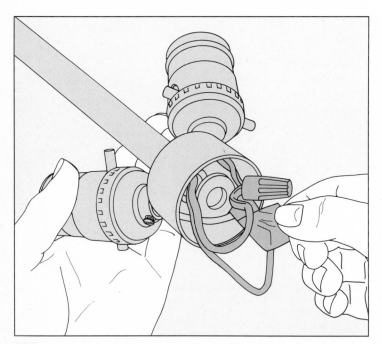

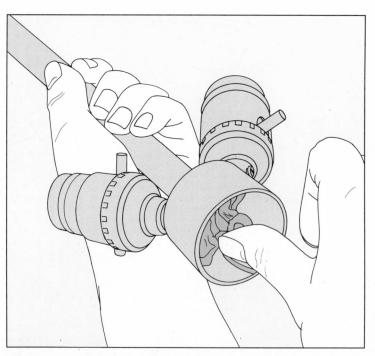

7 **Connecting the socket wires and lamp cord.** Twist the marked wire of the new lamp cord together with the two marked socket wires. Screw on a wire cap. Repeat the procedure for the unmarked wire of the new lamp cord and the two unmarked socket wires. As an extra precaution, secure the connections by wrapping the wire caps with electrical tape, as shown.

8 **Reassembling the lamp.** Fold the wires back into the top of the lamp, as shown. Replace the cover, shade and finial. Screw in the light bulbs and plug in the lamp to test it.

REWIRING A DESK LAMP

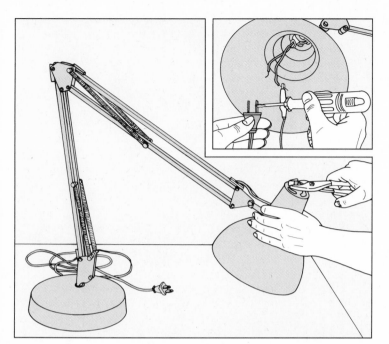

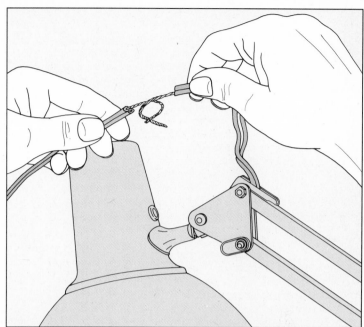

1 **Testing the cord and plug.** Unplug the lamp and remove the light bulb. Use pliers to unscrew the socket retaining ring *(above)*. Then push the cord in at the bottom of the lamp to gain some slack and pull the socket out of its ceramic or plastic insulating sleeve. Loosen the terminals with a screwdriver and disconnect the wires. Remove the socket and set it aside. If the plug is not polarized, place the alligator clip of the continuity tester on one wire end and touch the tester probe first to one plug prong *(inset)*, then the other. The tester should light for only one prong. Repeat the test for the other wire end. If the plug is polarized (one prong is wider than the other), place the alligator clip on the unmarked wire end and touch the tester probe to the narrow prong. Repeat the test for the marked wire end and wide prong. The tester should light for both tests. If not, tighten the connections and test again. If the cord and plug fail the second test, replace them *(next step)*.

2 **Splicing the old and new cords.** Untie the Underwriters' knot, if any, in the old cord, then pull the cord through the hole in the lamp shade so that its wire ends are exposed at the top of the upper arm. To create a splice that is thin enough to feed through the channel of the lamp arm, hook together only one wire of the old and new cords *(above)*. Secure the splice with electrical tape.

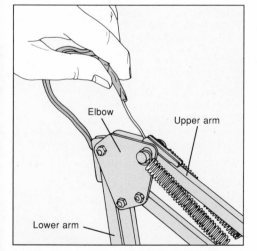

Elbow

Upper arm

Lower arm

3 **Fishing the new cord.** Use the old cord to pull the new cord through the upper arm until the splice appears at the elbow *(above)*, then continue pulling the new cord through the lower arm to the base. Undo the splice and remove the old wire.

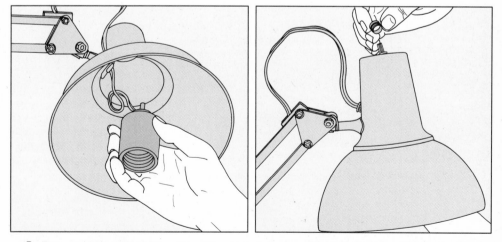

4 **Reassembling the lamp.** Feed the end of the new cord into the lamp shade, then part it and strip back 3/4 inch of insulation from each wire. Attach the ends to the socket terminals *(page 32, step 9)* and insert the socket into the insulating sleeve. Then pull the cord at the lamp elbow to create 3 or 4 inches of slack. Set the socket assembly inside the shade *(above, left)*, screw on the socket retaining ring *(above, right)* and tighten with pliers. Remove all but 2 inches of slack at the elbow by pulling the cord through from the table. Install a new plug *(page 36)*, then screw in a light bulb, plug in the lamp and test the repair.

HALOGEN LAMPS

Better lighting—for a price. Efficient, high-intensity halogen lighting may indeed be the "lighting of the future," as its manufacturers claim. Halogen lamps, named for the gas that fills their special bulbs, come in two basic styles. Table lamps are usually low-voltage and have a transformer, most often located in the base, that enables them to plug into a 120-volt outlet. Halogen floor lamps operate on regular voltage. But within these categories, there are many variations, making repair complicated. On some low-voltage models, you may be able to simply unscrew and remove the transformer; on others, the transformer is sealed in a case. Some models have fuses; others do not. Current may be carried from the transformer to the socket by low-voltage wire or conducted by insulated lamp supports. The switch may be attached by terminal screws or soldered to the lamp.

On average, halogen lamps are much more expensive than incandescent lamps. Although the bulbs are also costly, they can last up to twice as long as incandescent bulbs. Unlike standard bulbs, they give the same amount of light whether old or new.

Although halogen lamps are growing in popularity, they require extra caution even under normal use. Because halogen bulbs can reach much higher temperatures than incandescent bulbs, Underwriters Laboratories (UL) specifications require that some halogen bulbs have a protective covering. If the bulb cracks under high pressure, the guard will contain the shattered pieces. The guard also protects flammable materials—and hands—from touching the hot bulb.

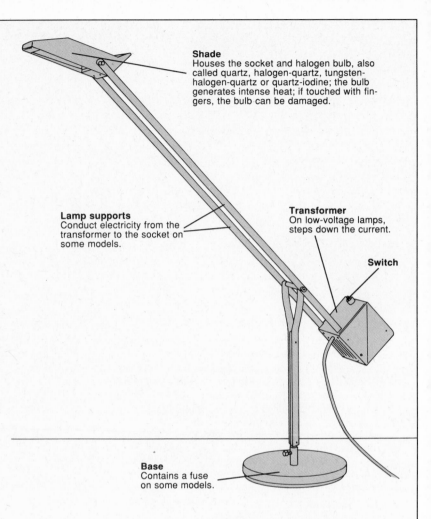

Shade
Houses the socket and halogen bulb, also called quartz, halogen-quartz, tungsten-halogen-quartz or quartz-iodine; the bulb generates intense heat; if touched with fingers, the bulb can be damaged.

Lamp supports
Conduct electricity from the transformer to the socket on some models.

Transformer
On low-voltage lamps, steps down the current.

Switch

Base
Contains a fuse on some models.

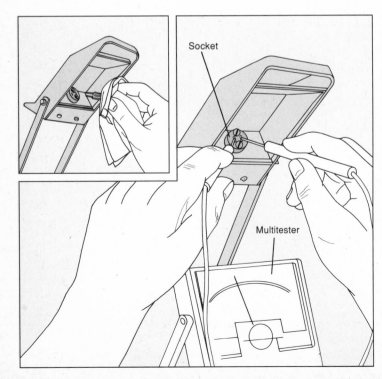

Socket

Multitester

Locating the problem. First perform the same routine checks as for incandescent lamps. Make sure the lamp is plugged in correctly, check the outlet by plugging another lamp into it, and look for a blown fuse or tripped circuit breaker. Then unplug the lamp and, being careful not to touch the bulb with your fingers, use a clean cloth to remove it from the lamp *(inset)*. Set a multitester to the AC scale that is just above the lamp's voltage requirement. For example, test a low-voltage lamp by setting the meter to the 50-volt scale. Touch the probes to the two socket terminals, as shown. If the meter shows the required voltage reading (in this case, 12 volts) the lamp is working properly and the bulb should be replaced. Install a new bulb using a clean cloth. If you accidentally touch the bulb, use rubbing alcohol on a cloth to clean the spot (oily deposits may damage the bulb when it gets hot).

When the multitester needle does not move, or shows a reading lower than the lamp's voltage requirement, the problem could be a faulty transformer (on low-voltage lamps), a burned fuse (on lamps with fuses), a broken cord (most common on lamps with flexible necks) or a melted or broken socket. Since models vary greatly and replacement parts can be difficult to find, take your lamp to a dealer for repairs.

LIGHTING FIXTURES

Lighting fixtures illuminate halls and stairways, brighten dim corners and spotlight treasured paintings or curios. Incandescent fixtures are still the most common type for home use, although fluorescent fixtures offer a practical alternative for kitchens and workrooms.

When a fixture doesn't light, first check the bulb. If it must be replaced, choose from the many shapes and colors on the market as shown on page 28. To repair the fixture, you will have to take it down from the wall or ceiling. Disconnecting and reconnecting the wiring is relatively simple; taking down and remounting the fixture, especially if it is heavy or has older mounting hardware, can be hard work.

Simple ceiling and wall fixtures are mounted directly to an electrical box. Heavier fixtures, such as chandeliers, must be secured with a mounting strap; a threaded nipple secures the strap to the stem of the fixture. Older fixtures may be supported by a threaded stud mounted to the back of the box. The versatile track fixture clips onto its channel; to disengage it, simply turn the lever on the stem. Some fixtures are recessed in the ceiling and must be pried out with care.

Carefully follow standard safety procedures when working on a fixture. Always flip off the wall switch and turn off power to the circuit by removing the fuse or tripping the circuit breaker *(page 18)*. Use a sturdy ladder to reach an overhead fixture and have a helper ready when taking down a heavy chandelier. Once you have exposed the wire connections, use a voltage tester to confirm that the power is off.

When reconnecting a fixture, pay special attention to the color coding of the wires. Fixture leads are black or unmarked (hot) and white or marked with a ridge, stripe or threaded tracer in the strands (neutral). Protect against shock by polarizing the fixture *(page 25)*; connect the hot wire to the brass terminal on the socket and to the black house wire, and connect the neutral wire to the silver terminal on the socket and to the white house wire. Replacement wire should be the same gauge as the old wire, and its insulation rated to withstand high temperatures.

Ceiling fans are a recent addition to the home fixture family. They come with a lighting fixture, or a light can be added later. For ceiling fan repairs, see page 56.

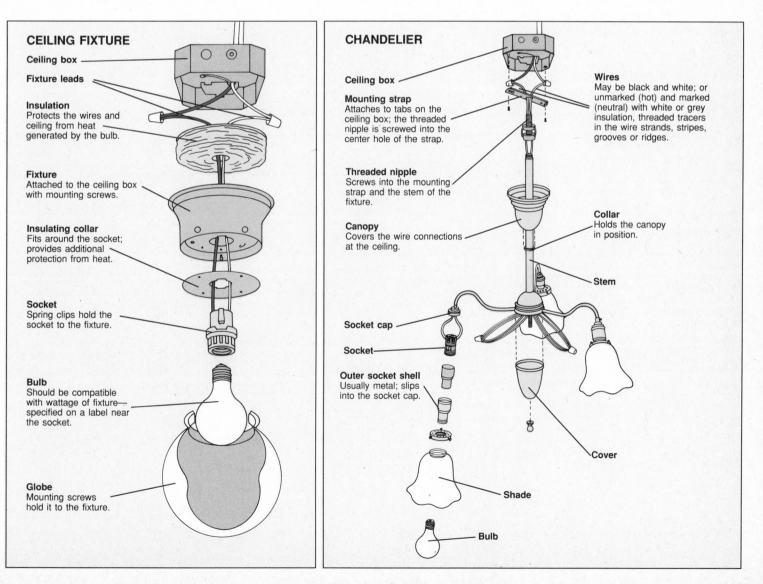

CEILING FIXTURE

Ceiling box

Fixture leads

Insulation
Protects the wires and ceiling from heat generated by the bulb.

Fixture
Attached to the ceiling box with mounting screws.

Insulating collar
Fits around the socket; provides additional protection from heat.

Socket
Spring clips hold the socket to the fixture.

Bulb
Should be compatible with wattage of fixture—specified on a label near the socket.

Globe
Mounting screws hold it to the fixture.

CHANDELIER

Ceiling box

Mounting strap
Attaches to tabs on the ceiling box; the threaded nipple is screwed into the center hole of the strap.

Threaded nipple
Screws into the mounting strap and the stem of the fixture.

Canopy
Covers the wire connections at the ceiling.

Socket cap

Socket

Outer socket shell
Usually metal; slips into the socket cap.

Wires
May be black and white; or unmarked (hot) and marked (neutral) with white or grey insulation, threaded tracers in the wire strands, stripes, grooves or ridges.

Collar
Holds the canopy in position.

Stem

Cover

Shade

Bulb

TROUBLESHOOTING GUIDE

SYMPTOM	POSSIBLE CAUSE	PROCEDURE
CEILING FIXTURE		
Bulb flickers or does not light	Bulb loose or burned out	Tighten or replace bulb
	No power to fixture	Replace fuse or reset circuit breaker *(p. 18)* □○
	Wall switch faulty	Check wall switch *(p. 70)*
	Socket contact dirty or bent too far down	Clean or bend up tab *(p. 45)* □○
	Socket and wires faulty	Test and replace socket and wires *(p. 45)* ▣○
	Socket or switch faulty (pull-chain socket)	Test socket and switch and replace socket *(p. 47)* ▣○
WALL FIXTURE		
Bulb flickers or does not light	Bulb loose or burned out	Tighten or replace bulb
	No power to fixture	Replace fuse or reset circuit breaker *(p. 18)* □○
	Wall switch faulty	Check wall switch *(p. 70)*
	Socket contact dirty or bent too far down	Clean or bend up tab *(p. 48)* □○
	Socket and wires faulty	Test and replace socket and wires *(p. 48)* ▣○
CHANDELIER		
One bulb (or more) flickers or does not light	Bulb loose or burned out	Tighten or replace bulb
	Socket contact dirty or bent too far down	Clean or bend up tab *(p. 49)* □○
	Socket faulty	Test and replace socket *(p. 49)* ▣◕
	Socket wires faulty	Test and replace socket wires *(p. 49)* ▣◕
	Wires in stem faulty	Test and replace wires in stem *(p. 49)* ▣◕
Entire fixture does not light	No power to fixture	Replace fuse or reset circuit breaker *(p. 18)* □○
	Wall switch faulty	Check wall switch *(p. 70)*
TRACK LIGHTING		
Bulb flickers or does not light	Bulb loose or burned out	Tighten or replace bulb
	Socket contact dirty or bent too far down	Clean or bend up tab *(p. 52)* □○
	Track contacts dirty	Clean track contacts *(p. 52)* □○
	Socket and wires faulty	Test and replace socket and wires *(p. 52)* ▣◕
All fixtures on track do not light	No power to track	Replace fuse or reset circuit breaker *(p. 18)* □○
	Wall switch faulty	Check wall switch *(p. 70)*
RECESSED FIXTURE		
Bulb flickers or does not light	Bulb loose or burned out	Tighten or replace bulb
	No power to fixture	Replace fuse or reset circuit breaker *(p. 18)* □○
	Wall switch faulty	Check wall switch *(p. 70)*
	Socket contact dirty or bent too far down	Clean or bend up tab *(p. 54)* □○
	Socket faulty	Test and replace socket *(p. 54)* ▣◕
	Socket wires faulty	Test and replace wires *(p. 54)* ▣◕

DEGREE OF DIFFICULTY: □ Easy ▣ Moderate ■ Complex
ESTIMATED TIME: ○ Less than 1 hour ◕ 1 to 3 hours ● Over 3 hours

SERVICING CEILING FIXTURES

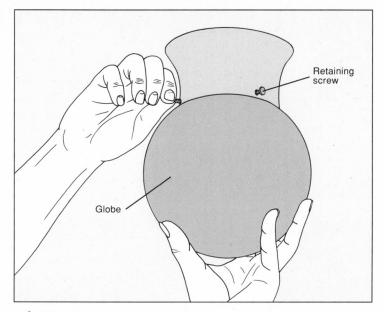

Retaining screw

Globe

1 **Dismounting the fixture.** Flip off the wall switch. Loosen any retaining screws holding the globe to the fixture, as shown, then remove the globe. Tighten a loose bulb or replace a burned-out bulb with one of the same wattage. Flip on the switch. If the fixture lights, reinstall the globe. If it doesn't, turn off power to the fixture by removing the fuse or tripping the circuit breaker *(page 18)*. Use a screwdriver to remove the mounting screws holding the fixture to the ceiling box.

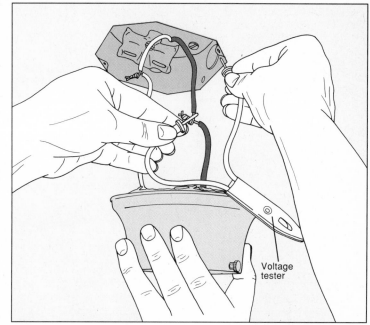

Voltage tester

2 **Testing for voltage.** Have someone hold the fixture, then twist off the wire caps to expose the connections, taking care not to touch any bare wire ends. Use a voltage tester to confirm that the power is off by touching one probe to the black wire connection and the other first to the grounded metal box *(above)*, then to the white wire connection. Then test between the white wire connection and the box. The tester should not glow in any test. If it does, return to the service panel and turn off power to the correct circuit. When the power is confirmed off, untwist the connections and take down the fixture.

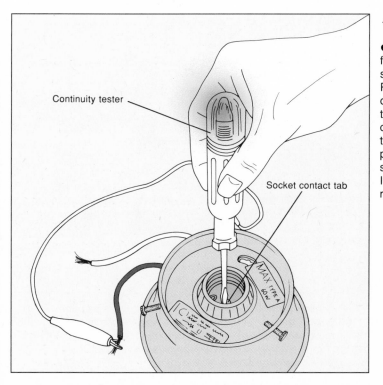

Continuity tester

Socket contact tab

3 **Testing the socket.** Pull away the insulation, if any, to expose the base of the socket. Scrape off any corrosion from the socket contact tab and pry it up slightly to improve contact with the bulb. Place the alligator clip of a continuity tester on the bare end of the black wire from the fixture and touch the tester probe to the socket contact tab, as shown. Then place the alligator clip on the white wire end and touch the probe to the threaded metal tube of the socket. The tester should light in both tests. If it doesn't, the socket and wires must be replaced.

SERVICING CEILING FIXTURES (continued)

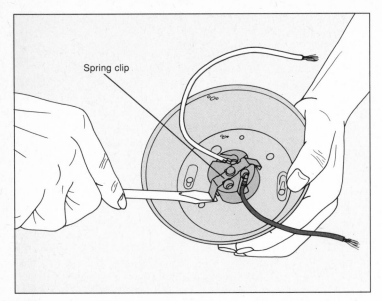

4 **Removing the old socket.** To free the socket from the fixture, insert a screwdriver under the spring clip on the socket and press on the clip, as shown. On other models, you may have to unscrew the two parts of the socket to remove it. Pull the socket and wires out of the fixture. Take the old socket with you to buy a compatible replacement with preattached wires.

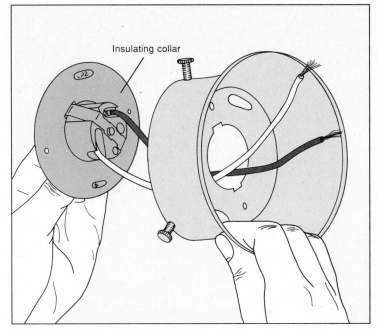

5 **Installing the new socket.** Thread the socket wires through the insulating collar, if any, then through the fixture (above). If the new socket has spring clips, as on the model above, push the socket into the fixture until the clips snap into place. If the replacement is a two-part, screw-type socket, fit the bottom half through the back of the fixture and screw the top half onto it from the front.

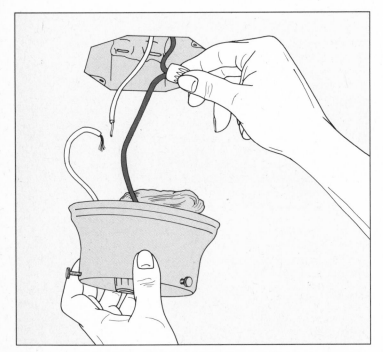

6 **Connecting the wiring.** Strip back the wires inside the box to improve the connections (page 141). Twist together the black fixture lead with the black wire from the ceiling box, and the white fixture lead with the white wire from the box, then screw a wire cap onto each connection (above).

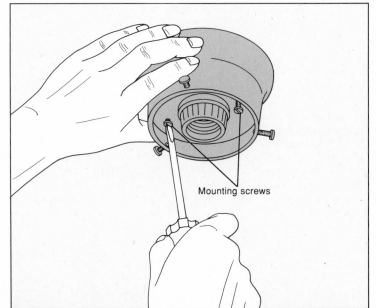

7 **Mounting the fixture.** Gently fold the wires into the ceiling box. Position the fixture so that its mounting slots align with the mounting holes on the ceiling box. Insert the mounting screws and tighten them (above). Screw in the bulb, then set the globe in place and tighten its retaining screws.

REPLACING A PULL-CHAIN SOCKET

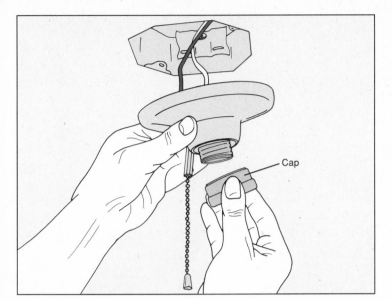

Cap

1 Access to the socket. The ceiling fixture shown above is usually made of porcelain and often found in garages, basements and closets. It has no wire leads; it is simply a housing for a socket that connects directly to the house wires. If the switch breaks or the socket fails, simply replace the socket. Turn off power to the fixture by removing the fuse or tripping the circuit breaker *(page 18)*. Take out the bulb, then loosen the screws that hold the fixture to the ceiling. Unscrew the cap from the socket *(above)*, then lower the fixture. Do not touch the socket or any metal parts until you have confirmed that the power is off *(next step)*.

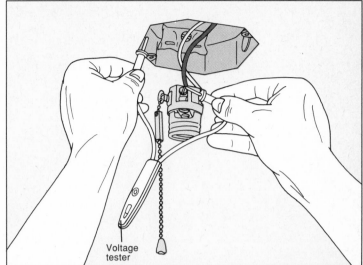

Voltage tester

2 Testing for voltage. Use a voltage tester to confirm that the power is off by touching one probe to the brass socket terminal and the other first to the grounded metal box *(above)*, then to the silver terminal screw. Then test from the silver terminal screw to the grounded box. The tester should not glow in any test. If it does, return to the service panel and turn off power to the correct circuit.

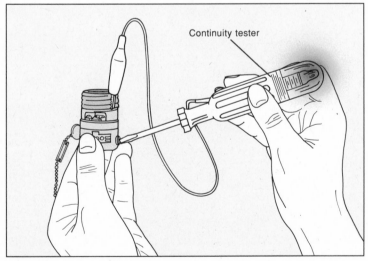

Continuity tester

3 Testing the socket and switch. With the power confirmed off, disconnect the wires from the terminal screws and take down the socket. Use a continuity tester to check the socket and switch. Place the alligator clip on the threaded metal tube and touch the tester probe to the silver terminal screw, as shown. The tester should light. Then place the alligator clip on the brass terminal screw and touch the tester probe to the socket contact tab. Pull the switch. The tester should light when the switch is in one position and not when it is in the other. If the socket fails either test, it should be replaced.

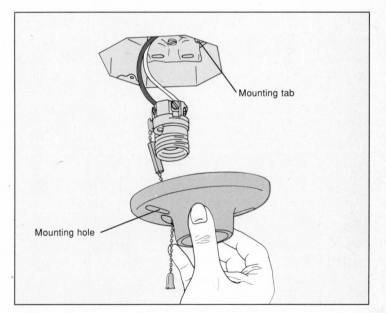

Mounting tab

Mounting hole

4 Replacing the socket. Take the old socket with you to buy a compatible replacement. Strip back the wires inside the box to improve the connections *(page 141)*. Connect the black wire to the brass terminal screw, then connect the white wire to the silver terminal screw. Gently fold the wires into the ceiling box. Thread the pull chain through its hole in the fixture *(above)*, then position the fixture so that its mounting holes align with the mounting tabs in the ceiling box. Insert the mounting screws and fasten them tightly with a screwdriver, reattach the cap and screw in the bulb. Turn on the power.

SERVICING WALL FIXTURES

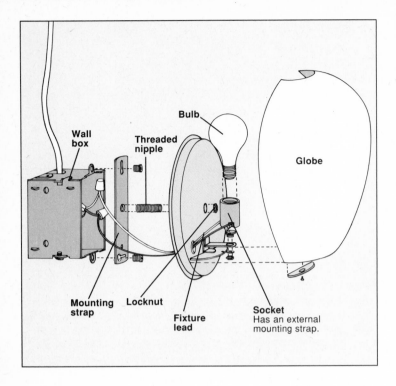

Wall box

Threaded nipple

Bulb

Globe

Mounting strap

Locknut

Fixture lead

Socket
Has an external mounting strap.

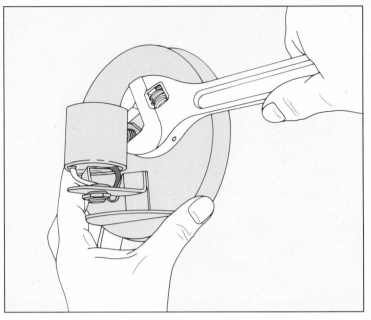

1 **Removing the fixture.** Flip off the wall switch and remove the globe. Tighten a loose bulb or replace a burned-out bulb with one of the same wattage. Flip on the switch. If the fixture lights, reinstall the globe. If it doesn't, turn off power to the fixture by removing the fuse or tripping the circuit breaker *(page 18)*. Unscrew the bulb. Use an adjustable wrench to loosen the locknut holding the fixture to the threaded nipple *(above)*, then pull the fixture away from the wall.

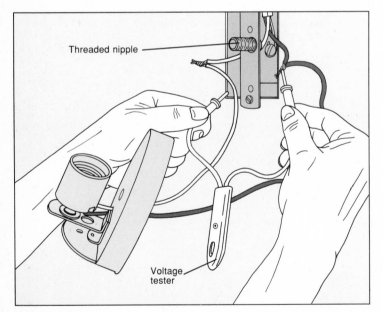

Threaded nipple

Voltage tester

2 **Testing for voltage.** Unscrew the wire caps from the black and white wire connections, taking care not to touch any exposed wire ends. Have someone support the fixture if it is heavy. Use a voltage tester to confirm that the power is off by touching one probe to the black wire connection and the other first to the grounded metal box *(above)*, then to the white wire connection. Then test between the white wire connection and the grounded box. The tester should not glow in any test. If it does, return to the service panel and turn off power to the forrect circuit. When the power is confirmed off, untwist the connections and take down the fixture.

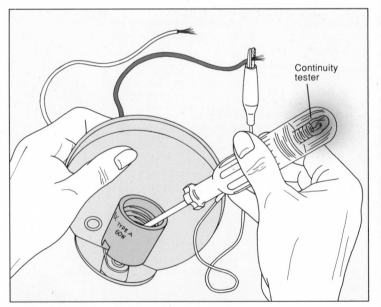

Continuity tester

3 **Testing the socket and wires.** Scrape off any corrosion from the socket contact tab and pry it up slightly to improve contact with the bulb. Place the alligator clip of a continuity tester on the black wire end and touch the tester probe to the socket contact tab, as shown. Then place the alligator clip on the white wire end and touch the probe to the threaded metal tube of the socket. The tester should light in both tests. If not, replace the socket and wires.

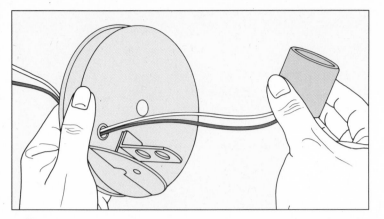

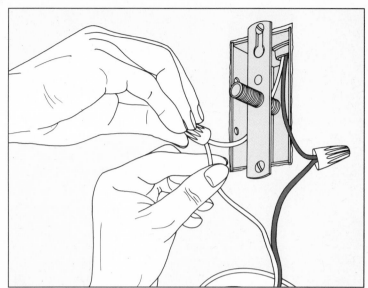

4 **Replacing the socket and wires.** Loosen the mounting screw holding the socket to the fixture, then pull the socket and the wires free of the fixture *(above)*. Buy a compatible socket with preattached wires or a socket with terminal screws and two pieces of wire the same gauge, length and color as the old ones. To connect the wires to the socket terminals, strip back the insulation *(page 138)*, twist the strands together and curl them into a hook, then loop the black wire end around the brass terminal screw and the white wire end around the silver terminal screw and tighten the connections. Thread the wires back through the hole in the fixture and tighten the mounting screw to secure the socket to the fixture.

5 **Reinstalling the fixture.** Strip back the wires inside the box to improve the connections *(page 141)*. Twist together the black lead from the fixture with the black wire in the box and screw on a wire cap. Join the white wires in the same way *(above)*. Gently fold the wires back into the box. Fit the fixture over the threaded nipple and position it against the wall, then slip on the locknut that holds the fixture to the nipple and tighten it. Screw in the bulb and replace the globe. Turn on the power.

SERVICING CHANDELIERS

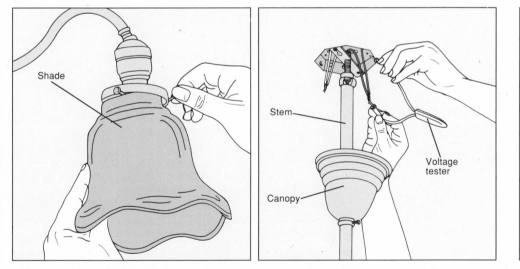

Shade

Stem

Canopy

Voltage tester

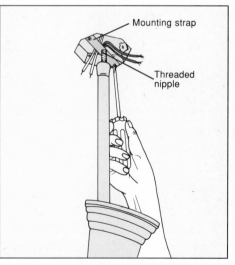

Mounting strap

Threaded nipple

1 **Testing for voltage.** After confirming that the problem is not a loose or burned-out bulb, flip off the wall switch and unscrew the light bulbs, then take off the shades *(above, left)* and carefully set them aside. Turn off power to the chandelier by removing the fuse or tripping the circuit breaker *(page 18)*. Loosen any screws holding the canopy in place against the ceiling and slide it down the stem to expose the ceiling box. Unscrew the wire caps from the connections, taking care not to touch any exposed wire ends. Use a voltage tester to confirm that the power is off by touching one of the tester probes to the black wire connection and the other probe first to the grounded metal box *(above, right)*, then to the white wire connection. Then test between the white wire connection and the grounded metal box. The tester should not glow in any test. If it does, return to the service panel and turn off power to the correct circuit.

2 **Dismounting the chandelier.** When the power is confirmed off, untwist the wire connections. Before dismounting the chandelier, use masking tape to tag the faulty part. Have a helper support the chandelier while you unscrew the mounting strap, as shown, and gently lower the fixture from the ceiling.

SERVICING CHANDELIERS (continued)

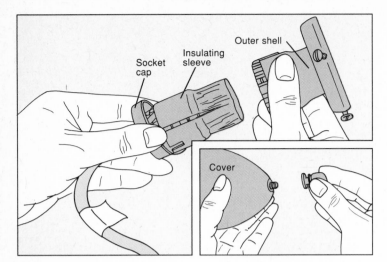

3 **Access to the socket.** To provide enough slack at the tagged sockets, remove the cover at the base of the fixture *(inset)*, pull out the wires and disconnect them. At the socket, press in with your thumb where the word "PRESS" is stamped and lift away the outer shell *(above)*. Then pull off the socket insulating sleeve. Pull the socket away from the socket cap so that at least 1 inch of wire is drawn out with it.

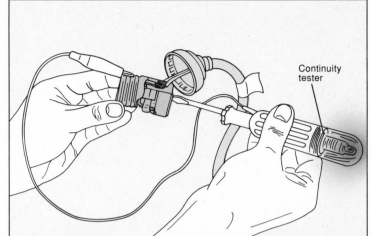

4 **Testing the socket.** Scrape any corrosion from the socket contact tab and pry it up slightly to improve contact with the bulb. To test the socket, place the alligator clip of a continuity tester on the threaded metal tube of the socket and touch the tester probe to the silver terminal screw *(above)*. Then place the alligator clip on the brass terminal screw and touch the tester probe to the socket contact tab. The tester should light in both tests. If not, loosen the socket terminal screws, detach the wires, remove the socket and install a replacement *(next step)*. If the socket tests OK, test the socket wires *(step 6)*.

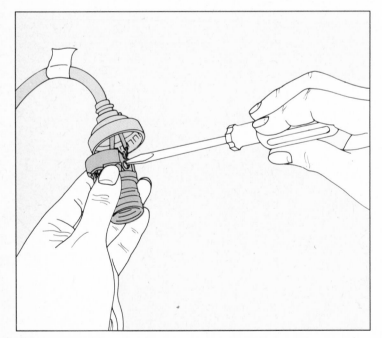

5 **Replacing the socket.** Buy a compatible socket with an insulating sleeve, an outer shell, and a socket cap. Replace only the faulty parts—in this case, the socket and insulating sleeve. Begin by removing the outer shell and insulating sleeve. Hook the marked wire around the silver socket terminal and tighten the connection, making sure there are no stray strands. Next, hook the unmarked wire around the brass socket terminal *(above)*. Slip the new insulating sleeve over the new socket, then push the outer shell into the cap until it snaps into place. At the base of the fixture, twist together the unmarked wires, then twist together the marked wires. Screw a wire cap onto each connection. Remount the chandelier *(step 10)*.

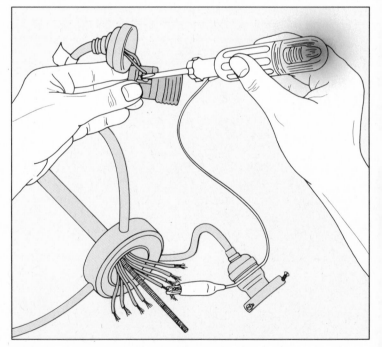

6 **Testing the socket wires.** Place the alligator clip of the continuity tester on the marked socket wire at the base of the fixture and touch the tester probe to the silver socket terminal, as shown. Then place the alligator clip on the unmarked socket wire and touch the tester probe to the brass socket terminal. The tester should light in both tests. If not, loosen the socket terminal screws, detach the wires, remove the socket and replace the socket wires *(next step)*. If the socket wires are good, go to step 8 to test the wires in the stem.

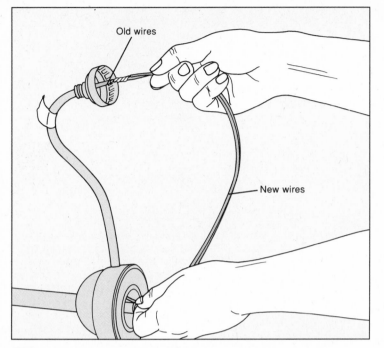

Old wires

New wires

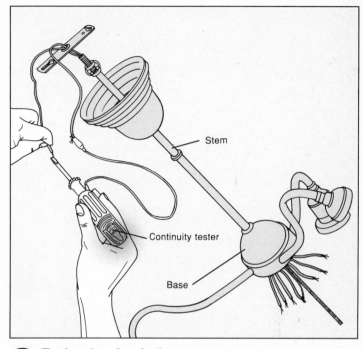

Stem

Continuity tester

Base

7 **Replacing the socket wires.** Buy wire of the same gauge and length as the old wire. Strip 3/4 inch of insulation from both ends of each wire *(page 138)*. Use the old wire to pull the new wire through the fixture arm. At the socket, hook the wires together and secure the connection with electrical tape, keeping the connection as thin as possible. Then, at the base of the fixture, pull out the old wires, bringing the new wires into the arm *(above)*. Connect the socket as in step 5. Remount the chandelier *(step 10)*.

8 **Testing the wires in the stem.** At the base of the fixture, twist together the two wires leading up the stem. At the top of the fixture, place the alligator clip of the continuity tester on one wire end and touch the tester probe to the other wire end, as shown. The tester should light. If it doesn't, replace the wires in the stem of the fixture *(next step)*.

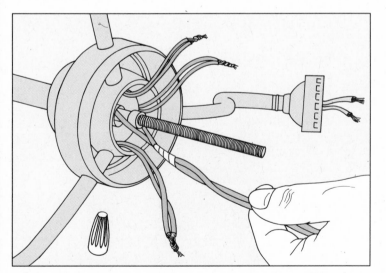

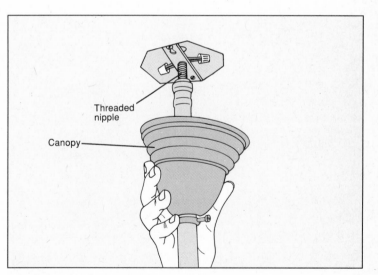

Threaded nipple

Canopy

9 **Rewiring the chandelier.** Buy wire of the same gauge and length as the old wire. Strip 3/4 inch of insulation from both ends of each wire *(page 138)*. Use the old wire to pull the new wire through the fixture stem. At the top of the stem, hook the wires together and secure the connection with electrical tape, keeping the connection as thin as possible. Then, at the base of the fixture, pull out the old wires, bringing the new wires through the stem *(above)*. Twist the marked socket wires together with the marked stem wire and screw on a wire cap. Repeat the procedure for the unmarked wires.

10 **Remounting the chandelier.** Fold the wire connections into the wiring compartment and screw on the cover. Strip back the wires inside the box to improve the connections *(page 141)*. While a helper supports the chandelier, tighten the mounting strap to the ceiling box. Twist together the marked fixture wire with the white wire in the box, and the unmarked fixture wire with the black wire in the box, then secure the connections with wire caps. Gently fold the wires into the box. Slide the canopy up the stem until it is flat against the ceiling *(above)* and tighten any screws to secure it. Screw in the bulbs and reattach the shades.

SERVICING TRACK FIXTURES

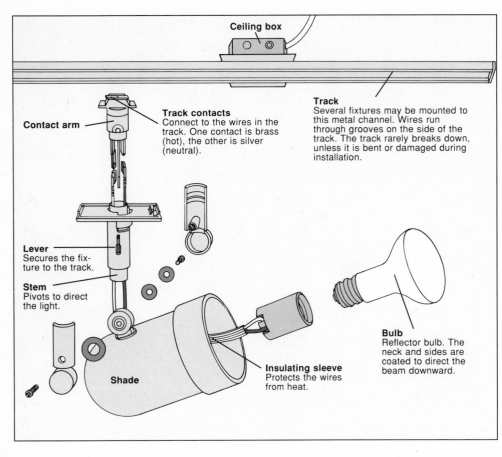

Ceiling box

Contact arm

Track contacts
Connect to the wires in the track. One contact is brass (hot), the other is silver (neutral).

Track
Several fixtures may be mounted to this metal channel. Wires run through grooves on the side of the track. The track rarely breaks down, unless it is bent or damaged during installation.

Lever
Secures the fixture to the track.

Stem
Pivots to direct the light.

Bulb
Reflector bulb. The neck and sides are coated to direct the beam downward.

Shade

Insulating sleeve
Protects the wires from heat.

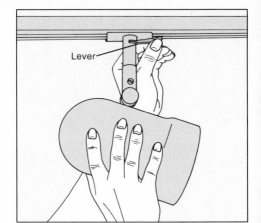

Lever

1 Cleaning the contacts. If the problem is not a loose or burned-out bulb, flip off the switch and turn off power to the track *(page 18)*. Allow the fixture to cool, then turn the lever to release the fixture from the track *(above)*. Do not touch any wires in the track. Use fine sandpaper to clean any dirt or corrosion from the track contacts on the fixture. Remove the bulb, then scrape any corrosion from the socket contact tab and pry it up slightly to improve contact with the bulb. Screw in the bulb, position the fixture in the track and flip on the switch. If the bulb still doesn't light, go to step 2.

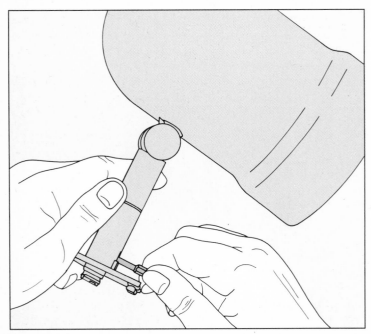

2 Access to the socket. Turn off power to the track and take down the fixture. To provide enough slack at the socket, unscrew the lever by hand *(above)*, then use a screwdriver to loosen the screw in the stem. Disassemble the stem. Remove the screws holding the socket to the shade and pull the socket free of its mounting.

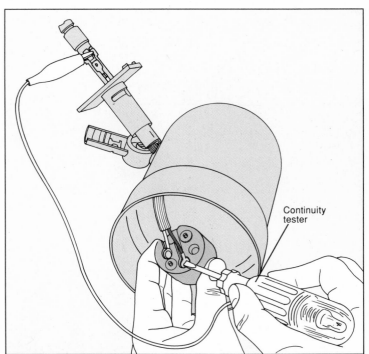

Continuity tester

3 Testing the socket. Place the alligator clip of a continuity tester on the brass track contact and touch the tester probe to the black wire connection at the socket terminal, as shown. Then place the alligator clip on the silver track contact and touch the tester probe to the white wire connection. The tester should light in both tests. If it doesn't, replace the socket and its wires *(next step)*.

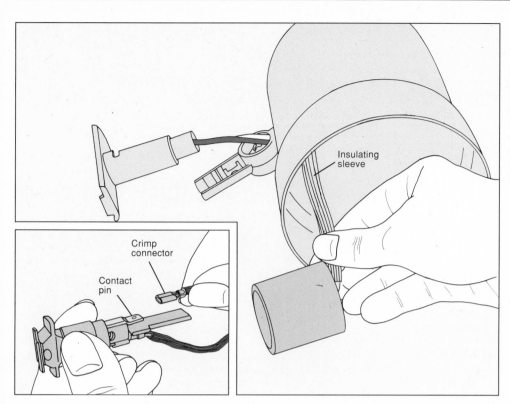

4 **Removing the old socket.** To release the wire connections at the base of the stem, pull the crimp connectors free of the contact pins *(inset)*. Then pull the socket out of the shade, as shown, bringing with it the wires and insulating sleeve. Slip the wires out of the insulating sleeve and set the sleeve aside; do not discard it. Take the socket and wires with you to buy a compatible socket with preattached wires. Also buy two crimp connectors the same size and shape as the old ones.

Insulating sleeve

Crimp connector

Contact pin

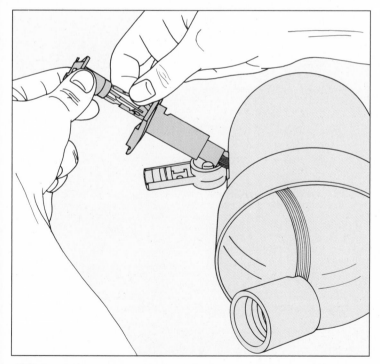

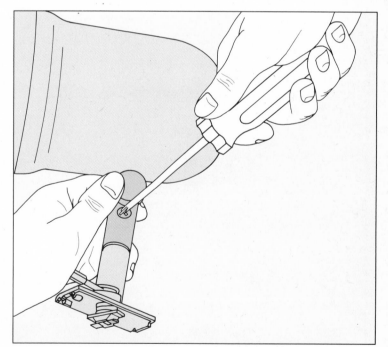

5 **Installing the new socket.** Slip the insulating sleeve over the new wires, then thread the wires through the hole in the shade and through both parts of the stem. Strip 1/4 inch of insulation from the wire ends *(page 138)*, then twist the strands tightly and push each wire end into a crimp connector. Use a multipurpose tool *(page 132)* to tighten the connectors. Next, push the crimp connector attached to the black wire onto the brass contact pin and the crimp connector attached to the white wire onto the silver contact pin *(above)*.

6 **Reassembling the fixture.** Set the socket in the shade and tighten the screws to secure it. Screw the lever into the stem by hand, then reassemble the stem and screw it together *(above)*. To mount the fixture, fit the stem in the track and turn the lever a quarter-turn. Do not touch any wires in the track. Screw in the bulb, turn on the power and flip on the switch.

SERVICING RECESSED FIXTURES

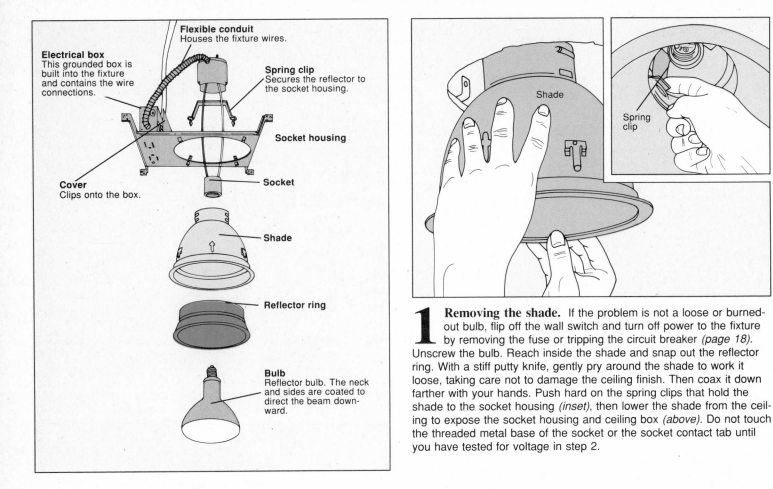

Electrical box
This grounded box is built into the fixture and contains the wire connections.

Flexible conduit
Houses the fixture wires.

Spring clip
Secures the reflector to the socket housing.

Socket housing

Cover
Clips onto the box.

Socket

Shade

Reflector ring

Bulb
Reflector bulb. The neck and sides are coated to direct the beam downward.

Shade

Spring clip

1 Removing the shade. If the problem is not a loose or burned-out bulb, flip off the wall switch and turn off power to the fixture by removing the fuse or tripping the circuit breaker *(page 18)*. Unscrew the bulb. Reach inside the shade and snap out the reflector ring. With a stiff putty knife, gently pry around the shade to work it loose, taking care not to damage the ceiling finish. Then coax it down farther with your hands. Push hard on the spring clips that hold the shade to the socket housing *(inset)*, then lower the shade from the ceiling to expose the socket housing and ceiling box *(above)*. Do not touch the threaded metal base of the socket or the socket contact tab until you have tested for voltage in step 2.

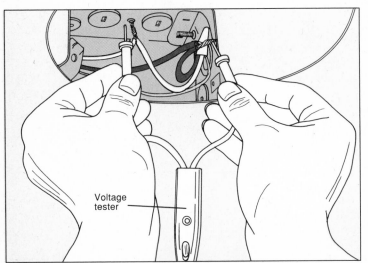

Voltage tester

2 Testing for voltage. Push aside the socket housing to gain access to the ceiling box. Unclip the cover from the box, then twist off the wire caps to expose the connections, taking care not to touch any bare wire ends. Use a voltage tester to confirm that the power is off by touching one probe to the black wire connection and the other probe first to the grounded metal box *(above)*, then to the white wire connection. Then test between the white wire connection and the box. The tester should not glow in any test. If it does, return to the service panel and turn off power to the correct circuit.

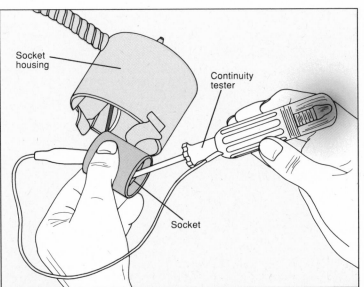

Socket housing

Continuity tester

Socket

3 Testing the socket. Scrape away any corrosion from the socket contact tab and pry the tab up slightly to improve contact with the bulb. Unscrew the socket from its mounting, then pull it out slightly to expose the connections at its base. Place the alligator clip of a continuity tester on the white wire connection at the base of the socket and touch the tester probe to the threaded metal tube, as shown. The tester should light. If it doesn't, replace the socket *(next step)*. If the socket is good, test the wires *(step 5)*.

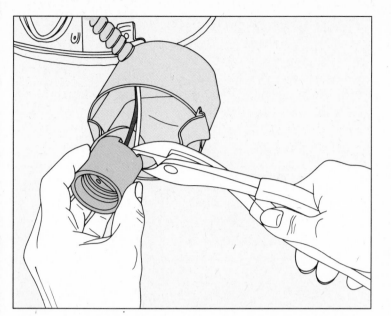

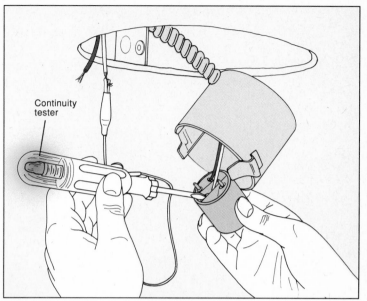

4 **Replacing the socket.** With diagonal-cutting pliers, cut the wires as close to the socket as possible *(above)*. Buy a replacement socket with terminal screws rather than preattached wires. Strip 3/4 inch of insulation from each wire end, twist the ends between your fingers, then loop the black wire end around the brass terminal screw and the white wire end around the silver terminal screw. Fit the socket back into its housing and tighten the mounting screws.

5 **Testing the socket wires.** Untwist the wire connections in the electrical box. Place the alligator clip of a continuity tester on the white socket wire and touch the tester probe to the white wire connection at the socket *(above)*. Then place the alligator clip on the black socket wire and touch the tester probe to the black wire connection at the socket. The tester should light in both tests. If it doesn't, buy a compatible socket with terminal screws and two pieces of wire the same gauge, length and color as the old ones. The wires should also have high-temperature insulation.

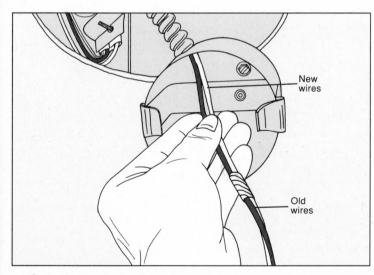

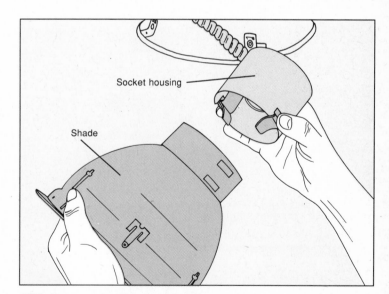

6 **Replacing the socket wires.** Use the old wires to pull the new wires through the flexible conduit. In the electrical box, temporarily splice the old and new wires and secure the connection with electrical tape. Then pull the old wires through the socket housing, bringing the new wires with them and leaving 8 inches of new wire in the electrical box. Disconnect the old wire and discard it. Twist together the black wire end with the black wire in the box, then the white wire end with the white wire in the box, and screw a wire cap onto each connection. Install the socket as in step 4.

7 **Remounting the fixture.** Gently fold the wires into the electrical box and snap the cover on the box. Hold down the spring clips and push the shade into the socket housing, as shown. Push the shade into place, making sure it is flush with the ceiling. Screw in the bulb and turn on the power.

CEILING FANS

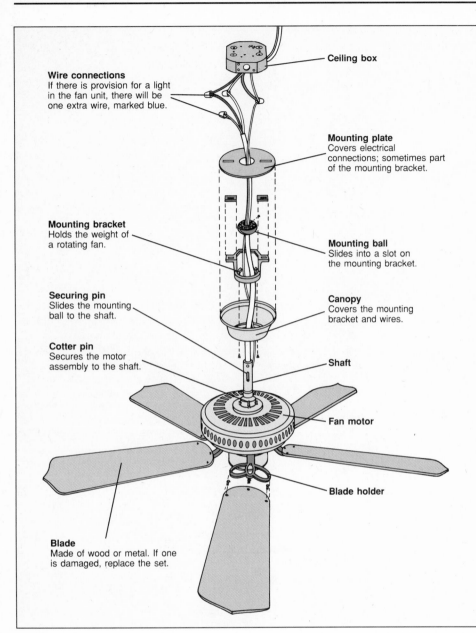

Wire connections
If there is provision for a light in the fan unit, there will be one extra wire, marked blue.

Ceiling box

Mounting plate
Covers electrical connections; sometimes part of the mounting bracket.

Mounting bracket
Holds the weight of a rotating fan.

Mounting ball
Slides into a slot on the mounting bracket.

Securing pin
Slides the mounting ball to the shaft.

Canopy
Covers the mounting bracket and wires.

Cotter pin
Secures the motor assembly to the shaft.

Shaft

Fan motor

Blade holder

Blade
Made of wood or metal. If one is damaged, replace the set.

Ceiling fans produce a cooling breeze in the summer and circulate trapped warm air in the winter. The unit is essentially an electric motor with blades attached to it, controlled by a variable-speed wall switch. Many ceiling fans also have a light fixture; others can have lights added later.

To start a stalled fan, make sure that the power is off and the wall switch is set at zero or turned off, then stand on a stepladder and slowly rotate the fan blades a few turns. Turn on the power and gradually turn the switch up to test.

Wobble and excessive vibration can result from worn pins that connect the shaft to the motor assembly and mounting ball. Replace these pins if they are faulty. Also check the fan blades for signs of damage or warpage. Replace the entire set, since they must weigh within 2 or 3 grams of each other for the fan to rotate evenly.

If you suspect that the motor has failed, take down the fan and test it. In most cases, the motor of the ceiling fan is a sealed unit and must be professionally serviced. If a ceiling fan does have to be replaced, the size of blades, fan speed and energy consumption are factors to consider in the purchase.

Turn off the power and test to confirm that the power is off before touching any bare wire ends. To take down the heavy unit, stand on a sturdy ladder and have a helper support the fixture while you disconnect it.

TROUBLESHOOTING GUIDE

SYMPTOM	POSSIBLE CAUSE	PROCEDURE
Fan does not operate	No power to the circuit	Replace fuse or reset circuit breaker *(p. 18)* □○
	Wall switch faulty	Test and replace wall switch as you would a dimmer switch *(p. 70)*
	Motor faulty	Rotate blades with hands Test motor and take for service *(p. 57)* ◖●▲
Fan wobbles while operating	Fan blades not properly balanced	Replace all blades
	Mounting loose	Tighten bolts and screws and replace worn pins at the top and bottom of the shaft *(p. 57)* ◖●
	Motor faulty	Test motor and take for service *(p. 57)* ◖●▲

DEGREE OF DIFFICULTY:	□ Easy ◖ Moderate ■ Complex
ESTIMATED TIME:	○ Less than 1 hour ◖ 1 to 3 hours ● Over 3 hours

▲ Multitester required

SERVICING CEILING FANS

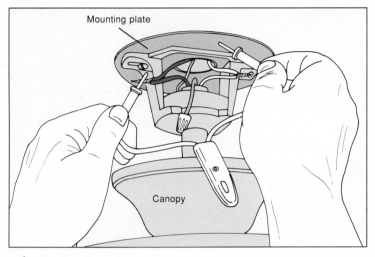

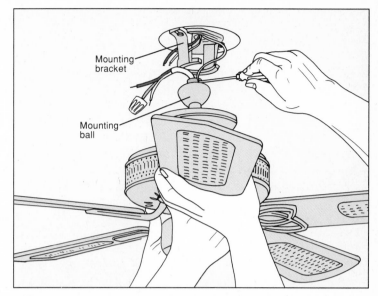

1 Testing for voltage. Turn off the switch, then turn off power to the fan by removing the fuse or tripping the circuit breaker *(page 18)*. Loosen the screws holding the canopy and slide it down the shaft to expose the mounting bracket. Pull the black and white wire connections and the blue lighting wire, if any, out of the ceiling box. Leave the grounding wire connections in the box. Unscrew the wire caps from the black and white wire connections, taking care not to touch any exposed wire ends. Use a voltage tester to confirm that the power is off by touching one probe to the grounded metal mounting plate and the other to the black and white wire connections in turn *(above)*. The tester should not glow in either test. If it does, return to the service panel and turn off power to the correct circuit.

2 Taking down the fan. Ceiling fans weigh as much as 50 pounds; have a helper support the unit while you dismount it. Begin by disconnecting the wires, then raise the fan, slip the mounting ball out of the mounting bracket, and hand the fan to the helper. Push the mounting ball down until you see the green grounding wire attached to the grounding screw at the top of the shaft. Unscrew it from the shaft, as shown, then lower the fan.

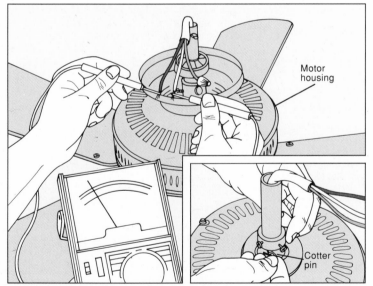

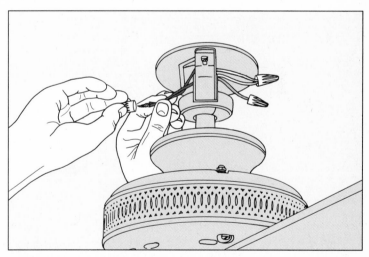

3 Testing the motor and servicing the fan. Set a multitester *(page 135)* to the RX1 scale to test the motor. Touch one tester probe to the black lead and the other to the white lead, as shown. The tester should register between 20 and 30 ohms. If there is no reading, the motor is faulty; take the fan for servicing or buy a replacement. If the motor passes the test, inspect the mounting. Pull out the cotter pin that secures the motor housing to the fan shaft and inspect it for wear *(inset)*. Replace a worn cotter pin with a new one. Tighten the screws connecting the motor to the shaft. Also check the pin that holds the mounting ball in place and, if it appears worn, replace it. Inspect the blades; if any are broken or warped, replace the set. Check the blade holders to see if they are bent and tighten any loose blade screws.

4 Reinstalling the fan. Have the helper hold the fan while you reconnect the grounding wire to the top of the shaft. Slide the mounting ball up the shaft. Slip the mounting ball into its slot in the mounting bracket. Twist together the black fan wire with the black wire in the box and the white fan wire with the white wire in the box and screw on wire caps, as shown. Gently fold the connections back into the box. Slide the canopy up the shaft and screw it into place. To adjust a canopy that cuts into the ceiling when the fan rotates, reposition it, leaving a gap of 1/4 inch between the canopy and ceiling, or insert washers between the mounting bracket and the mounting plate or the mounting plate and the ceiling. Turn on the power and turn on the wall switch.

FLUORESCENT LIGHTING

Fluorescent tubes can shed six times more light per watt of power and last five times longer than incandescent bulbs. This makes fluorescent lighting an economical choice for kitchens, workshops and areas where bright light is needed for long periods of time. The color of light cast, once a cold blue-white, can now be almost any desired shade.

The working parts common to all fluorescent fixtures and lamps are tubes, sockets and ballasts; starters or trigger switches are also found on some models. Tubes *(page 60)* can be straight to fit rectangular fixtures, circular to fit circline fixtures, and U-shaped to fit one-socket fixtures. Replacement tubes must match the ballast and socket wattage.

Sockets come in a variety of sizes and shapes and are held in place by the fixture frame, cover plate or a socket bracket. If a socket is broken or its metal contacts are bent or corroded, replace it *(page 62)*.

The heart of a fluorescent fixture is the ballast, a transformer-like device that initially boosts the incoming current to start the tubes, then reduces the voltage to the level required for continuous lighting. The ballast can last up to 12 years and is generally the most expensive part to replace; when it does fail, you might want to consider replacing the fixture altogether. One sign of an old or faulty ballast is a black resin that drips out of the ballast casing, caused by overheating.

There are two basic types of fluorescent fixtures for household use: the starter type (sometimes called pre-heat) and the rapid-start. Starter-type fixtures, usually older units of 30 watts or less, have two circuits: the first provides the initial surge of voltage; the second supplies current for continuous lighting. Rapid-start models do not require a starter for the initial power surge. They are quicker to light but consume slightly more power.

When a fluorescent fixture is not working properly, first look for a blown fuse or tripped circuit breaker at the main service panel *(page 18)*. Next check for a burned-out tube, a faulty wall switch *(page 70)* or, on starter-type fixtures, a broken starter. You may have to replace the socket, the ballast or the entire fixture.

A fluorescent fixture may be surface-mounted on a wall or ceiling in a kitchen or bathroom, or recessed in a suspended ceiling of a recreation room. When replacing the fixture, choose one that is mounted in the same way as the old one.

Fluorescent lamps work on the same basic principle as fluorescent fixtures. One popular model, the trigger-switch lamp *(page 64)*, resembles a starter-type fixture. The switch performs the same function as the starter; pressing it for several seconds provides the necessary surge of voltage to start the lamp. A more recent generation of fluorescent lamps includes a compact, one-socket model *(page 65)*, identified by its small, U-shaped tube. The components of this lamp—the switch, sockets and small, lightweight ballast—are found in the lamp head.

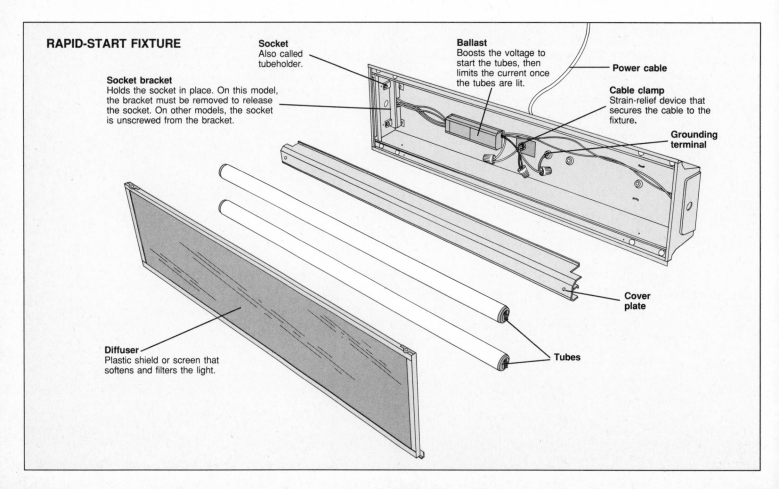

RAPID-START FIXTURE

Socket
Also called tubeholder.

Socket bracket
Holds the socket in place. On this model, the bracket must be removed to release the socket. On other models, the socket is unscrewed from the bracket.

Ballast
Boosts the voltage to start the tubes, then limits the current once the tubes are lit.

Power cable

Cable clamp
Strain-relief device that secures the cable to the fixture.

Grounding terminal

Cover plate

Tubes

Diffuser
Plastic shield or screen that softens and filters the light.

TROUBLESHOOTING GUIDE

SYMPTOM	POSSIBLE CAUSE	PROCEDURE
Fixture doesn't light or ends of tube glow but center doesn't	No power to fixture	Replace fuse or reset circuit breaker *(p. 18)* □○
	Tube burned out	Replace tube *(p. 61)* □○
	Wall switch faulty	Check wall switch *(p. 70)*
	Tube pins making poor contact in sockets	Reseat tube *(p. 61)* □○; replace sockets *(p. 62)* ▪○
	Starter faulty	Replace starter *(p. 61)* □○
	Ballast faulty	Replace ballast *(p. 63)* ▪◗; replace fixture *(recessed or surface-mounted, p. 63; mounted in suspended ceiling, p. 64)* ▪◗
	Grounding connections faulty (rapid-start)	Tighten grounding connections *(p. 62)* □○
Tube blinks, flickers or is slow to light	Tube damp or dirty	Wipe tube with window cleaner and dry thoroughly
	Short, frequent use; tube requires time to stabilize	Leave light on for longer periods
	Starter faulty	Replace starter *(p. 61)* □○
	Tube pins making poor contact in sockets	Reseat tube *(p. 61)* □○; replace sockets *(p. 62)* ▪◗
	Ballast faulty	Replace ballast *(p. 63)* ▪◗; replace fixture *(recessed or surface-mounted, p. 63; mounted in suspended ceiling, p. 64)* ▪◗
	Location too cold	Install ballast rated for cold temperatures
Fixture hums	Ballast vibrating	Remount ballast *(p. 63)* □○
	Ballast faulty	Replace ballast *(p. 63)* ▪◗; replace fixture *(recessed or surface-mounted, p. 63; mounted in suspended ceiling, p. 64)* ▪◗
Black resin seeps from fixture	Ballast faulty	Replace ballast *(p. 63)* ▪◗; replace fixture *(recessed or surface-mounted, p. 63; mounted in suspended ceiling, p. 64)* ▪◗
Desk lamp doesn't light or lights incompletely	Tube burned out	Replace tube *(trigger-switch, p. 64; one-socket, p. 65)* □○
	No power to lamp	Plug lamp into another outlet to test
	Switch faulty	Replace switch *(trigger-switch, p. 64; one-socket, p. 65)* ▪◗
	Tube pins making poor contact in sockets	Reseat tube *(p. 61)* □○; replace sockets *(trigger-switch, p. 64; one-socket, p. 65)* ▪○
	Ballast faulty	Replace ballast *(trigger-switch, p. 64; one-socket, p. 65)* ▪◗

DEGREE OF DIFFICULTY: □Easy ▪Moderate ▪Complex
ESTIMATED TIME: ○Less than 1 hour ◗1 to 3 hours ●Over 3 hours

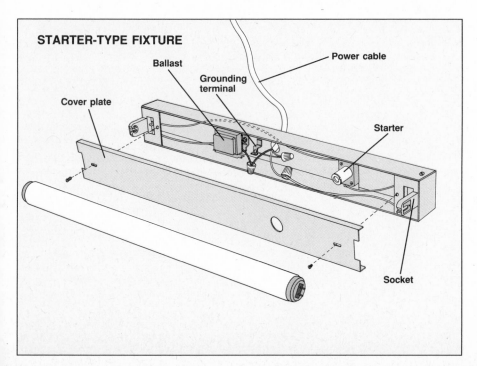

STARTER-TYPE FIXTURE

Power cable

Ballast

Grounding terminal

Cover plate

Starter

Socket

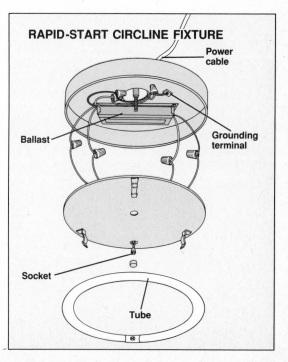

RAPID-START CIRCLINE FIXTURE

Power cable

Ballast

Grounding terminal

Socket

Tube

FLUORESCENT TUBES

Care of fluorescent tubes. Unlike incandescent bulbs, which light when current flows through the filament, a fluorescent tube lights when electricity charges a gas inside the tube, causing the phosphorus-coated inner surface to glow. Age, rapid switching, humidity, dirt and cold can create an electrochemical imbalance inside the tube and result in swirling, fluttering or blinking. Follow these steps to enhance the performance of the fluorescent tubes:

- Remove a dirty or damp tube, wipe it with a clean cloth dampened with window cleaner and dry it thoroughly.

- Leave a fluorescent light on for several hours rather than switching it on and off for occasional use. Life expectancy drops with increased switching.

- Allow a flickering tube, new or old, to stabilize by leaving it on for several hours.

- Do not expect a fluorescent fixture to perform well at temperatures below 50°F. Install a cold-rated ballast for unheated basements or garages. Shielded tubes are available if wind is a problem.

- Take care when handling a broken fluorescent tube; the mercury inside is poisonous and the glass can cause cuts and slivers.

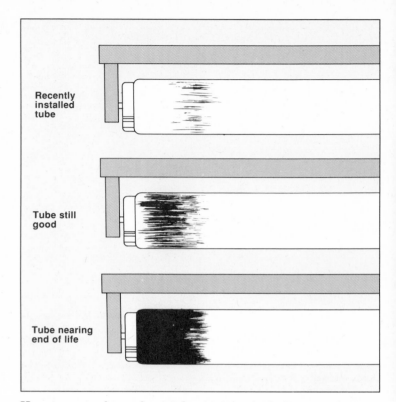

How to spot a burned-out tube. As the cathode filaments of a fluorescent tube wear away, they form black deposits at each end. Light discoloration at the ends indicates a nearly new tube, a wider band of gray is a normal mid-life condition and a blackened end means that the tube is almost finished. If only one end is blackened, reverse the tube to extend its life.

CHOOSING A TUBE

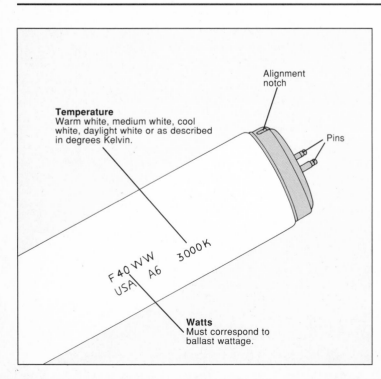

Alignment notch

Temperature
Warm white, medium white, cool white, daylight white or as described in degrees Kelvin.

Pins

F 40 WW   3000K
USA   A6

Watts
Must correspond to ballast wattage.

Fluorescent color and temperature. When changing a fluorescent tube, read the information printed at the end of the tube near the manufacturer's name. Be sure to buy a replacement of the same wattage and length as the old one. Proper length generally assures compatible wattage; if you are unsure, check the wattage requirement stamped on the ballast. Tubes identified as pre-heat are designed to be used in starter-type fixtures, but regular tubes can be used with only slightly less efficiency.

You can choose from a wide variety of temperatures and colors. Temperature ratings range from cool to warm and are measured in degrees Kelvin (or °K). Cool tubes (4000°K and up) produce the harsh factory light that has given fluorescents a bad name for home use. These tubes are the least expensive and most useful for task lighting where high visibility is needed. Warm tubes (3000°K or less) produce light comparable to incandescent bulbs. If in doubt, choose a medium tube (3000 to 4000°K) or, in double-tube fixtures, use one warm and one cool tube.

Color rendering describes the ability of a light source to illuminate objects. Sunlight is the standard, with a Color Rendering Index (CRI) of 100. Fluorescent grow tubes for plants rate in the 90's, while a standard warm white tube is about 50. The higher the CRI, the more expensive the tube.

REPLACING THE TUBE AND STARTER

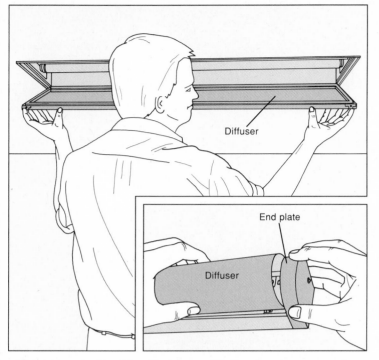

Diffuser

End plate

Diffuser

1 **Access to the tube.** Fluorescent fixtures often have a translucent cover, or diffuser, which must be removed to gain access to the tube. If the diffuser rests in a metal frame with pins on one side and clips on the other, slide in the clips to release one side and let the diffuser hang by the pins *(above)*. To release a diffuser held by end plates *(inset)*, pull one end plate out with one hand and pull the diffuser free with the other.

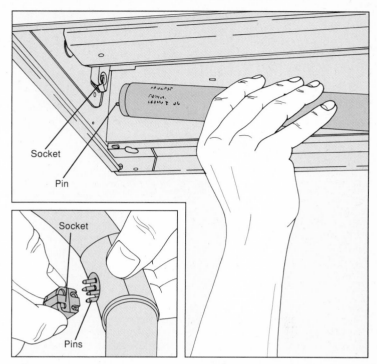

Socket

Pin

Socket

Pins

2 **Removing the tube.** Rotate the tube a quarter-turn in either direction and lower the pins from the sockets, as shown. To detach a circline tube, unplug the tube pins from the socket, then pull the tube free of its clips *(inset)*. To replace a starter, go to step 3. To replace a burned-out fluorescent tube, go to step 4.

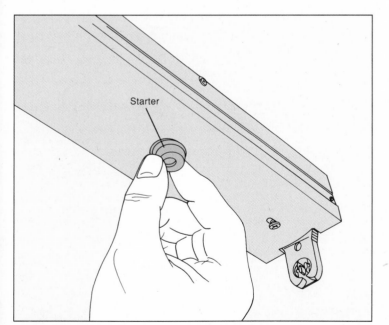

Starter

3 **Replacing the starter.** A defective starter is a likely cause of flickering or blinking and is much cheaper to replace than a tube. Push in the starter and twist it counterclockwise a quarter-turn *(above)*, then pull it out. Take it to a lighting store for a correct match; the wattage of the starter should correspond to that of the tube and ballast. Push in the new starter until its pins enter the slots, then turn clockwise until it clicks into place.

4 **Installing the tube.** Position the pins in the sockets *(above)*, then twist the tube a quarter-turn to seat it. The tube should fit snugly; if not, the socket may need replacement *(next page)*. Install a circline tube by lining up the pins with the holes in the socket and connecting them, then pushing the tube past the clips. Replace the diffuser.

SERVICING THE GROUNDING CONNECTIONS

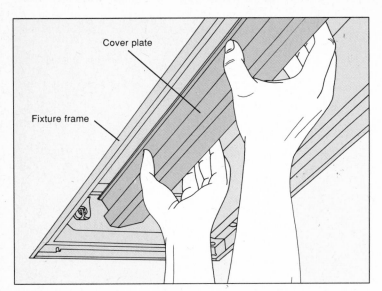

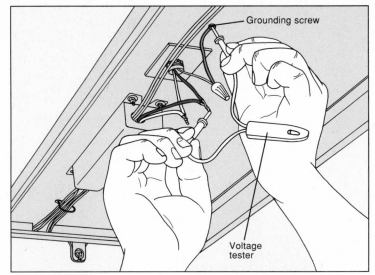

1 **Removing the cover plate.** A rapid-start fixture will not operate correctly if it is improperly grounded. To expose the wire connections, turn off power to the fixture by removing the fuse or tripping the circuit breaker *(page 18)*, then remove the diffuser and tubes *(page 61)*. Squeeze the cover plate to release it from the frame *(above)*, exposing the fixture's wiring.

2 **Testing for voltage and checking the grounding.** Unscrew the wire caps from the black and white leads, taking care not to touch any bare wire ends. Touch one probe of a voltage tester to the grounding screw and the other probe first to the black wire ends *(above)*, then the white wire ends. Then test between the black and white wire ends. The tester should not glow. Inspect the connections at the grounding screw and pigtail and tighten them if necessary. Remount the cover plate and install the tubes *(page 61)*. Turn on the power and flip on the switch. If the fixture lights, put back the diffuser.

REPLACING THE SOCKET

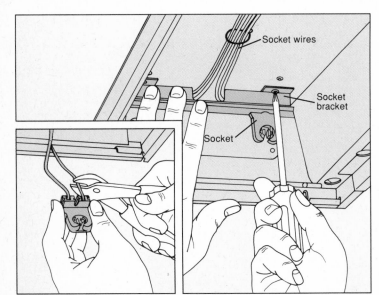

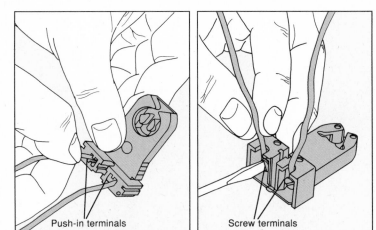

1 **Removing the socket.** Turn off the power *(page 18)*, remove the diffuser and tube *(page 61)*, take off the cover plate and test for voltage *(steps 1 and 2, above)*. Access the socket connections by unscrewing the socket bracket, as shown. To disconnect a socket with push-in terminals, insert a paperclip into each terminal slot and pull out the wire. To disconnect a socket with screw terminals, loosen the screws and unhook the wires. Sockets with preattached wires cannot be disconnected; cut the wires close to the socket *(inset)*, leaving long leads on the ballast.

2 **Replacing the socket.** Take the socket (and its removable bracket, if any) to a lighting store for the best possible match. Push each ballast wire end into a terminal *(above, left)* and tug gently to test for a secure connection. For screw terminals, curl each ballast wire end around a terminal *(above, right)* and tighten. Screw the socket and bracket, if any, back in place. Then reassemble the fixture, turn on the power and flip on the wall switch. If the fixture lights, replace the diffuser. If it doesn't, check the ballast *(page 63)*.

REPLACING THE BALLAST

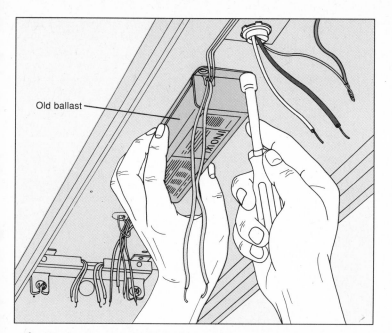

Old ballast

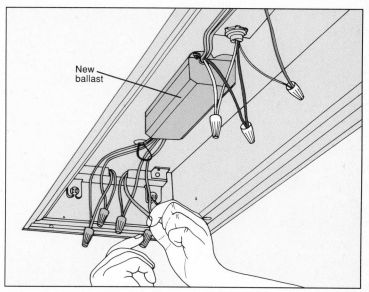

New ballast

1 **Removing the old ballast.** To test the ballast, flip off the wall switch and turn off power to the fixture at the service panel *(page 18)*. Expose the ballast by removing the diffuser and tubes *(page 61)*, then the cover plate *(page 62)*. Twist the wire caps off the black and white leads and grounding pigtail, taking care not to touch any exposed wire ends. Use a voltage tester to confirm that the power is off *(page 62)*. Release the sockets from their brackets, if any, then disconnect the ballast from both sockets. For sockets with preattached wires, cut the wires near the ballast, leaving a long lead on each socket. Since the ballast is the heaviest part of the fixture, it should be supported firmly with one hand while removing its mounting nuts or screws *(above)*. Take the old ballast with you for a correct match.

2 **Mounting the new ballast.** Set the new ballast in position with its black and white leads facing the incoming house wiring and hold it firmly while tightening the mounting nuts or screws. Connect the ballast to the sockets, twisting together the leads and screwing a wire cap on each connection, as shown. Set the sockets back in place, then screw or snap on the cover plate, making sure that the wires are not pinched. Install the tubes *(page 61)*, turn on the power and flip on the wall switch. If the ballast hums excessively, turn off the power and tighten the mounting nuts. Replace the diffuser and turn on the power.

REPLACING RECESSED OR SURFACE-MOUNTED FIXTURES

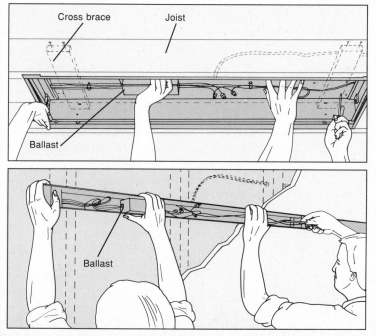

Cross brace   Joist

Ballast

Ballast

Flip off the wall switch and turn off power to the fixture at the service panel *(page 18)*. Remove the diffuser and tubes *(page 61)*, then the cover plate *(page 62)*. Twist the wire caps off the black and white leads and grounding pigtail, taking care not to touch any exposed wire ends, then use a voltage tester to confirm that the power is off *(page 62)*. If the power cable is attached to the fixture with a cable clamp, unscrew the locknut on the clamp. Recessed fixtures are held to cross braces between the joists with mounting screws *(left, top)*. Surface-mounted fixtures are usually attached directly to the ceiling or wall *(left, bottom)*. Have someone support the fixture and ballast while you remove the screws, then lower the fixture.

To install a new fixture, have someone hold it while you push the power cable into the unit through the knockout on the back plate. Drill new holes in the cross brace, if necessary, then secure the fixture by tightening the mounting screws at both ends of the fixture. Connect the black lead to the black house wire and the white lead to the white house wire. Then make a new grounding pigtail by connecting a jumper wire to the grounding terminal in the back of the fixture, and twisting it together with the bare grounding wire from the cable. Secure the connections with wire caps. Replace the cover plate, then install the tubes and diffuser.

REPLACING A FIXTURE IN A SUSPENDED CEILING

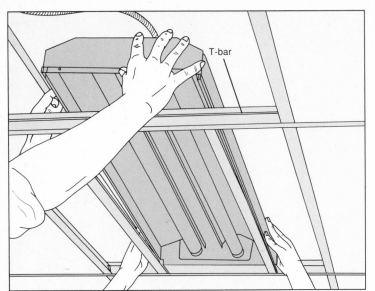

With power to the fixture turned off at the service panel *(page 18)*, remove the diffuser and tubes *(page 61)*, then the cover plate *(page 62)*. Twist the wire caps off the black and white leads and grounding pigtail, taking care not to touch any exposed wire ends, then use a voltage tester to confirm that power is off *(page 62)*. Unscrew the lock-nut on the cable clamp so that the cable can be pulled free from the fixture. Remove the ceiling tiles around the fixture, then reach in and detach any safety chains or straps securing the fixture. With a helper, lift, tilt and lower the fixture from the ceiling *(left)*. Buy a replacement fixture of the same dimensions.

Have someone hold the new fixture while you push the power cable through the knockout on the back plate and secure it by tightening the cable clamp. Set the fixture into the T-bar frame, then reattach any safety chains or straps. Connect the black lead to the black house wire and the white lead to the white house wire. Then make a new grounding pigtail by hooking a jumper wire around the grounding terminal in the back of the fixture and twisting it together with the bare grounding wire from the cable. Secure the connections with wire caps, screw or snap on the cover plate, then install the tubes and diffuser.

SERVICING TRIGGER-SWITCH LAMPS

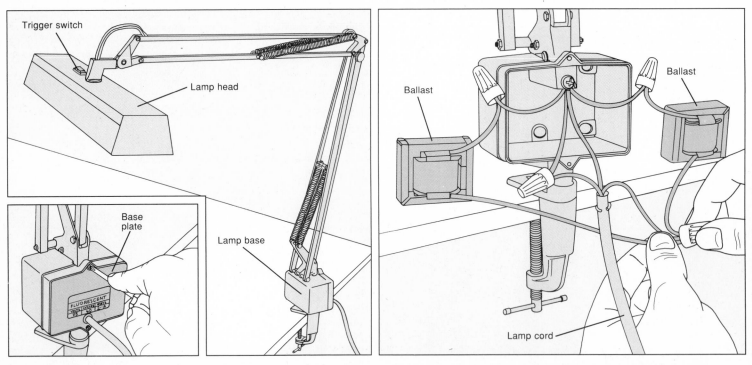

When a trigger-switch lamp *(above, left)* is not working correctly, switch it off and check the tube *(page 60)*; if it is burned out, install a compatible replacement. If the tube is good, unplug the lamp and look for visible damage in the plug and cord; replace them in the same way as for an incandescent lamp *(page 26)*. Also replace a broken or damaged socket *(page 62)*.

More often than not, the problem is with the ballast. Lamp ballasts are small and, because they are quite heavy, are usually located in the lamp base. When the lamp has two tubes, it often has a separate ballast for each. You can save time by replacing both ballasts while the lamp is disassembled. With the lamp unplugged, unscrew the base plate *(inset)* and slide it out of the way. Gently pull the wires out

of the base but do not detach them. Remove any mounting screws and place the ballasts beside the base. Twist off the wire caps *(above, right)*, and detach the two wires from one ballast at a time. If the wires are not color-coded, twist them together to identify them as a pair. Take the ballasts to the store for a correct match. To reassemble the lamp, twist each of the new ballast wires together with a lamp wire and screw on a wire cap. Set the ballasts and wiring back into the base, making sure that no wires are pinched.

If the problem is not the ballast, the trigger switch may be at fault. To access it, remove any snap-in or screw-on cover in the lamp head. Disconnect the wires from the three terminals and install a compatible replacement. Screw on the cover, plug in the lamp and test.

SERVICING ONE-SOCKET LAMPS

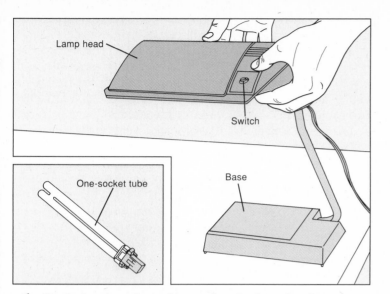

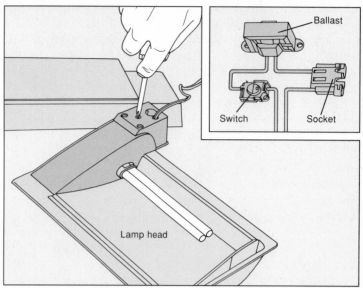

1 **Replacing the tube and removing the lamp head.** Most servicing of the compact, one-socket lamp takes place at the lamp head. To remove a burned-out tube, turn off the lamp, then grasp the tube as close to the base as possible, wiggle it back and forth and gently pull it free. Since the shape of the tube base changes with wattage, take the old tube with you to purchase a replacement. Holding the new tube at the base, push it firmly into the socket. If this doesn't solve the problem, unplug the lamp. If there is visible damage to the plug and cord, replace them as you would for an incandescent lamp *(page 26)*. If not, loosen any screws or clamps holding the lamp head on the stem, then slide it up and off *(above)*.

2 **Disassembling the lamp head.** Remove any screws from the component case in the lamp head, as shown, and pry off the cover with a screwdriver. Take note, as the lamp head is opened, of how the ballast, switch and socket are connected *(inset)*, as well as their positions in the lamp head. To replace the socket, go to the next step; to replace the switch or ballast, go to step 4.

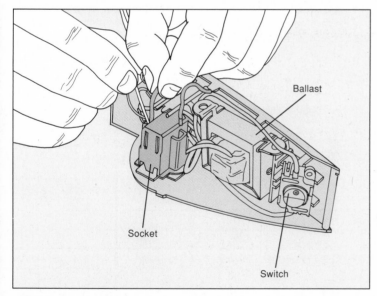

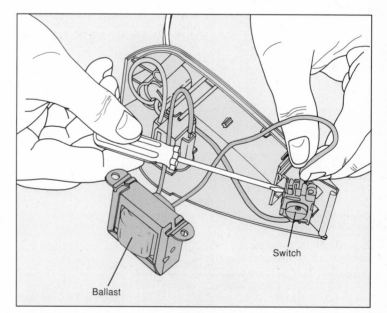

3 **Replacing the socket.** The single socket on this lamp has two push-in terminals. To disconnect the socket, push a pin or paperclip into the slot beside the wire and pull the wire free *(above)*. Then squeeze the two plastic prongs together and pull the socket out of the lamp. Buy a replacement socket of the same wattage and push it into place until the prongs catch, then test it with a light tug. Set the components back into the case, tuck in the wires, and reattach the cover plate. Return the assembled head to the lamp arm.

4 **Replacing the switch and ballast.** Disconnect the switch by detaching the wires at its two screw terminals. Disconnect the ballast by detaching one wire at the switch terminal and the other wire at the socket terminal. Take the old ballast or switch for a correct match. Reconnect the wires to the switch at the two screw terminals; reconnect one ballast wire to a switch terminal and the other to the socket terminal. Set the parts back into the lamp head, tuck in the wires, and reattach the cover plate. If the lamp still doesn't work, test the cord and plug as for an incandescent desk lamp *(page 26)*.

DOORBELLS AND CHIMES

Whether they chirp, chime, buzz or play the first few bars of "My Old Kentucky Home," all doorbell systems operate on low-voltage current supplied by a transformer. When your doorbell isn't operating, first check the push button, which takes the most wear and tear and is subject both to weather and constant use. Push buttons come in lighted or unlighted models, and can be surface-mounted or recessed in the wall or door frame. With only a screwdriver, you can easily take apart a push button, clean, dry and adjust the contacts. If the push button is badly worn or damaged, replace it.

The next likely candidate for repair is the transformer, usually located in the basement and attached to a junction box. The transformer taps into a house circuit and steps down the current to the low voltage required by the chime or bell. A chime usually takes a transformer rated at 16 volts, while a bell requires 10 to 20 volts. If you have to replace a faulty transformer, check the voltage and volt-ampere requirements of your bell or chime and buy a transformer that will deliver sufficient power.

Next, inspect the chime or bell for a loose connection and check the sounding components. While electronic door chimes are difficult to repair because of their complicated circuitry, mechanical chimes and bell units can usually be maintained by cleaning and lubricating their plungers. When a unit does fail, it is often simpler to install a replacement than attempt a repair. There are many sound units to choose from: mechanical chimes with an extended note for the front door and a short note for the back door, electronic chimes that can be programmed to play a variety of tunes, traditional bells that sound when a clapper strikes a metal gong, and buzzers.

The last resort in tracking down a doorbell problem is to inspect the small-diameter, low-voltage (16-, 18- or 20-gauge) bell wire that connects the components. If it is exposed, repair damaged wiring by cutting back and splicing it together with a wire cap. If the wire is behind the wall and you can't isolate the damaged portion, disconnect it and run new wire in its place *(page 102)*, or along baseboards and door trim.

Since chimes and bells operate on very low voltage, repairs to them present little danger of electric shock. However, to be sure, turn off the power to the doorbell system while working on the bell wires at the push button, chime and transformer. And, of course, when handling the connections between the transformer and the 120-volt household wiring, turn off the power and test to be sure that it is off.

TROUBLESHOOTING GUIDE

SYMPTOM	POSSIBLE CAUSE	PROCEDURE
Chime or bell unit does not operate	No power to system	Check for blown fuse or tripped circuit breaker *(p. 18)* □○
	Push-button contacts corroded	Clean contacts *(p. 67)* □○
	Push-button contacts bent too far down	Pry up contacts *(p. 67)* □○
	Button wiring broken	Splice on new wire *(p. 67)* □○
	Push button faulty	Test and replace push button *(p. 68)* □○
	Transformer connections loose	Tighten connections *(p. 68)* □○
	Transformer faulty	Test and replace transformer *(p. 68)* ◨●▲
	Chime or bell unit dirty	Clean unit *(p. 69)* □○
	Connections loose at chime or bell	Tighten connections *(p. 69)* □○
	Chime or bell unit faulty	Test and replace unit *(p. 69)* ◨●▲
	Wire broken along low-voltage line	Repair visible damage by cutting and splicing with wire caps □○ Rewire the doorbell system *(p. 102)* ◨●
Chime or bell rings constantly	Push-button contacts bent too far up	Bend contacts down *(p. 67)* □○
Chime fails to complete full cycle	Chime unit dirty	Clean and lubricate unit *(p. 69)* □○
	Transformer faulty	Test and replace transformer *(p. 68)* ◨●▲

DEGREE OF DIFFICULTY: □ Easy ◨ Moderate ■ Complex
ESTIMATED TIME: ○ Less than 1 hour ◖ 1 to 3 hours ● Over 3 hours   ▲ Multitester required

TYPES OF DOORBELLS AND CHIMES

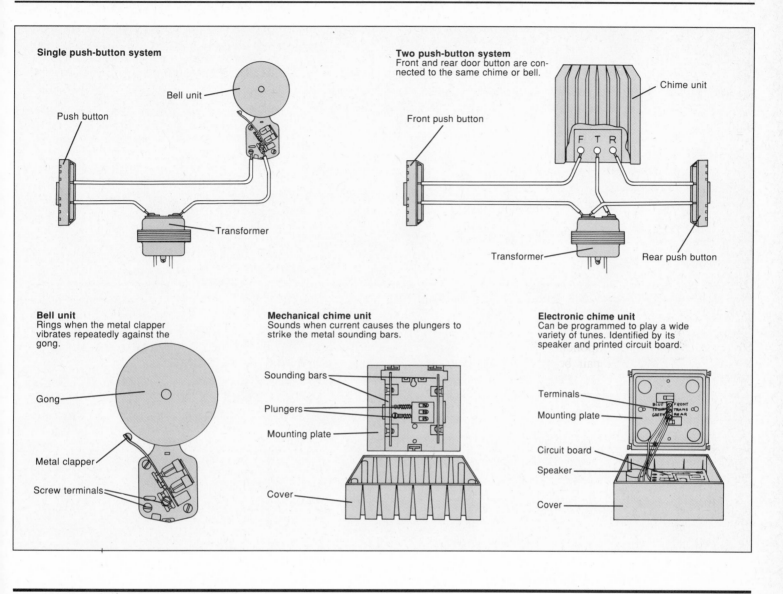

Single push-button system

Bell unit

Push button

Transformer

Two push-button system
Front and rear door button are connected to the same chime or bell.

Chime unit

Front push button

F T R

Transformer

Rear push button

Bell unit
Rings when the metal clapper vibrates repeatedly against the gong.

Gong

Metal clapper

Screw terminals

Mechanical chime unit
Sounds when current causes the plungers to strike the metal sounding bars.

Sounding bars

Plungers

Mounting plate

Cover

Electronic chime unit
Can be programmed to play a wide variety of tunes. Identified by its speaker and printed circuit board.

Terminals

Mounting plate

Circuit board

Speaker

Cover

SERVICING THE PUSH BUTTON

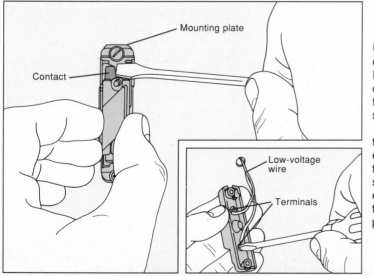

Mounting plate

Contact

Low-voltage wire

Terminals

1 **Servicing the push-button contacts and connections.** Turn off power to the doorbell by removing the fuse or tripping the circuit breaker *(page 18)*. Pry off or unscrew the push-button cover. Use fine sandpaper to gently clean the metal contacts. For a better connection, pry up the contacts slightly with a screwdriver, as shown. Replace the cover, turn on the power and press the button. If the chime or bell rings constantly, the contacts are bent up too far. Turn off the power, remove the push button and bend the contacts down slightly.

If the chime or bell does not ring at all, loosen any screws securing the mounting plate to the wall or door frame, then pull it forward to expose the wire connections. Loosen the terminal screws and unhook the wire connections. Cut back any breaks in the exposed wiring and splice the wire ends together with wire caps. Clip the exposed wire ends, strip the insulation *(page 141)* and secure the wire ends around the terminals *(inset)*. Reinstall the push button, turn on the power and press the button. If the chime or bell does not sound, go to step 2.

SERVICING THE PUSH BUTTON (continued)

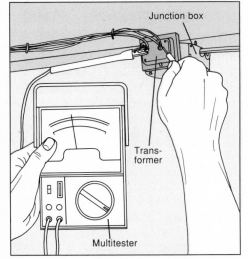

Stripped wire ends

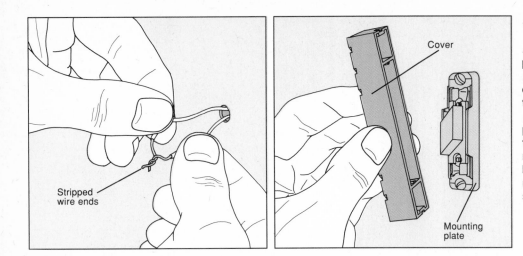

Cover

Mounting plate

2 Testing and replacing the push button. Remove the cover and mounting plate, disconnect the wires from the push-button terminals and twist them together *(far left)*. Then turn on the power. If the chime or bell does not sound, leave the wires twisted together and check the transformer *(below)*. If the chime or bell now rings, the push button needs to be replaced. To install the new push button, turn off the power, connect the wires to the terminal screws on the back of the new mounting plate, screw the mounting plate to the wall or door frame and snap on the cover *(near left)*.

TESTING AND REPLACING THE TRANSFORMER

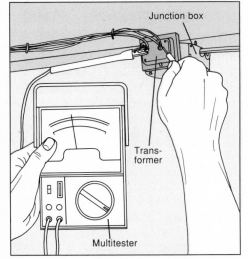

Junction box

Transformer

Multitester

1 Servicing the transformer. The transformer is usually attached to a junction box in the basement. The 120-volt connections are contained in the box; the low-voltage wires lead from the transformer to the doorbell. Turn off the power to the circuit *(page 18)*, tighten any loose connections at the low-voltage terminals, then turn on the power. If the wires are not already twisted together at the push button, have someone press the button. If the doorbell or chime does not sound, test the transformer. Set a multitester *(page 135)* to the ACV scale and turn the dial to the 50-volt range. With the power on, place one probe on each low-voltage terminal of the transformer, as shown. If the tester registers no voltage, service the 120-volt connections *(next step)*. If the tester reading corresponds to the voltage stamped on the side of the transformer (10 to 20 volts), check the chime or bell unit *(page 69)*.

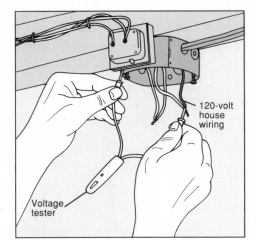

120-volt house wiring

Voltage tester

2 Servicing the 120-volt connections. Turn off power to the circuit by removing the fuse or tripping the circuit breaker *(page 18)*. Unscrew the junction box cover to expose the 120-volt connections. Unscrew the wire caps, taking care not to touch any exposed wire ends. Use a voltage tester to confirm that the power is off by touching one tester probe to the grounded metal box and the other probe to each exposed wire in turn *(above)*. Then test between the two wire connections. The tester should not glow. If it does, return to the service panel and turn off the power. Next, detach the wire ends and clean with fine sandpaper, or clip them back and strip off the insulation *(page 141)*, then reattach them. Turn on the power and have someone press the push button if its wires are not already twisted together. If the chime or bell sounds, turn off the power, screw on the wire caps, fold the wires back into the junction box and screw on its cover. If the bell or chime still does not work, replace the transformer.

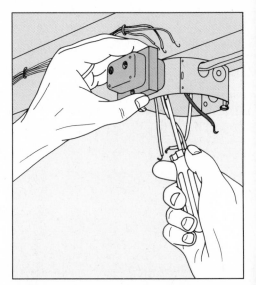

3 Replacing the transformer. With the power off, loosen the terminal screws on the low-voltage side and detach the doorbell wires. Then disconnect the transformer wires inside the junction box and unscrew the transformer from the box *(above)*. Buy a replacement transformer of the same voltage and volt-ampere rating. Thread the transformer leads through the knockout in the junction box, then twist one lead to the black house wire, the other to the white wire and secure each connection with a wire cap. As an added precaution, wrap electrical tape around the wire caps. Screw the transformer to the junction box. Attach the bell wires to the transformer terminals. Turn on the power and have someone press the push button if its wires are not already twisted together. If the doorbell works, replace the junction box cover. If it doesn't, service the chime or bell unit *(page 69)*.

SERVICING THE CHIME OR BELL UNIT

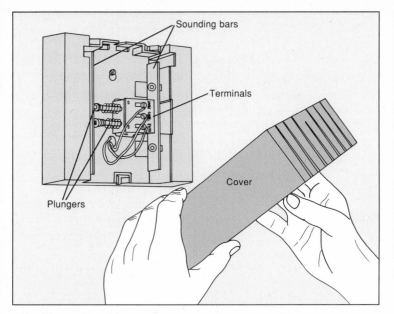

1 **Removing the cover.** Remove any screws holding the cover in place and pull it forward to reveal the sounding components. On mechanical chimes, the components are attached to the mounting plate on the wall *(above)*. On most electronic models, they are attached to the cover. Most chime units have three terminals, labelled FRONT, TRANS and REAR, connected to the front push button, transformer and rear push button respectively. Most bells have two terminals; one connects to the transformer, the other to the push button.

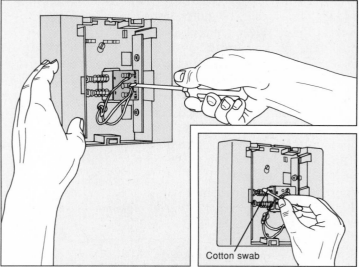

2 **Servicing the unit.** Turn off power to the circuit by removing the fuse or tripping the circuit breaker *(page 18)*. Clean or tighten any dirty or loose terminal connections, or cut back the wire ends and reattach them to the terminals, as shown. To clean mechanical chimes, blow away any dirt from between the plungers and sounding bars and clean the plungers with a cotton-tipped swab dipped in alcohol *(inset)*. Do not use lubricating oil, which may leave a sticky residue on moving parts. (Electronic chimes cannot be cleaned.) If a bell unit makes a dull sound, clean the gong and metal clapper in the same manner. Then turn on the power and have someone press the push button if its wires are not already twisted together. If the chime or bell still does not work, go to step 3.

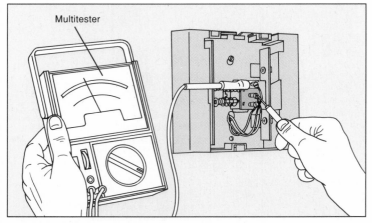

3 **Testing the unit.** Use a multitester *(page 135)* to verify that sufficient voltage is reaching the chime or bell unit. Check the reading specified on the low-voltage side of the transformer. Set the multitester to 50 volts on the ACV scale. For chime units, place one tester probe on the front button terminal and the other on the transformer terminal. Repeat the test for the rear button terminal, if there is one. For bell units, place one tester probe on each terminal. If the multitester reading is within 2 volts of the rating stamped on the transformer, the chime or bell unit is receiving sufficient voltage. This indicates that the unit is faulty and should be replaced *(next step)*. No reading indicates a break in the wiring between the front button and the transformer or chime, or between the back button and transformer or chime. With power off, rewire the doorbell system, fishing the wire where necessary *(page 102)*.

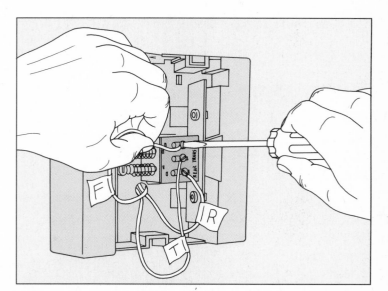

4 **Replacing the unit.** Turn off the power. Mark the wires with masking tape labelled FRONT, TRANS and REAR, then disconnect them. Remove the screws that hold the unit to the wall. Buy a replacement that matches the voltage and amperage ratings of the old transformer. Thread the wires through the mounting plate of the new unit. Hook the correct wire around each terminal, then tighten the screws, as shown. Screw the mounting plate to the wall, then snap or screw the cover in place. Turn on the power and press the push button, if its wires are not already twisted together.

WALL SWITCHES

With a few inexpensive tools such as an insulated screwdriver, voltage tester and continuity tester *(page 134)*, any wall switch can be safely replaced or restored to proper working order. The job can be as simple as cleaning corroded wire ends or tightening loose connections.

Switches found in most homes fall into three general categories, based on the number used to operate a single fixture: single-pole, three-way, and least common, four-way *(below)*.

When a fixture controlled by a wall switch does not work, do not service the switch immediately. First replace the light bulb, change the fuse or reset the circuit breaker at the main service panel *(page 18)*. If you suspect that the problem is in the circuit wiring, flip on the other switches and test a lamp in the outlets on that circuit. After eliminating these potential problems, look to the switch.

Although wall switches may last up to twenty years, heavy use can wear out the mechanism. Most moving parts are sealed inside the housing and cannot be accessed. When the mechanism fails, simply replace the switch.

When removing a wall switch from its box, do not be surprised by wiring variations. Hookups in switch boxes differ according to the number of terminals on the switch *(below)*, grounding variations *(page 73)*, the number of cables entering the box and the number of switches housed within it. The main trick to replacing a wall switch is simply to hook the correct wire around the correct terminal. Color coding on wires and terminals will help you do this. In some of the more complicated installations described on the pages that follow, you will be instructed to tag a wire with masking tape before disconnecting it to aid in reassembly.

The number of cables entering a switch box is determined by the location of the switch in the electrical circuit. A switch installed mid-circuit is called middle-of-the-run and has two (or more) cables in the box—one leading from the power source, the other going out to the next switch, outlet or fixture on the circuit. A switch installed at the end of the run is connected to the single cable in the box.

Manufacturers offer a wide range of wall switches for various locations and special needs. There are switches that turn a fixture on and off automatically, or set mood lighting. Some switches are fitted with a pilot light or oversize handle; others can be locked to protect dangerous power tools. For most, installation is as simple as putting in a single-pole switch. Be sure to read the markings stamped on the back of the switch and the mounting strap to ensure that you select a suitable replacement *(page 72)*.

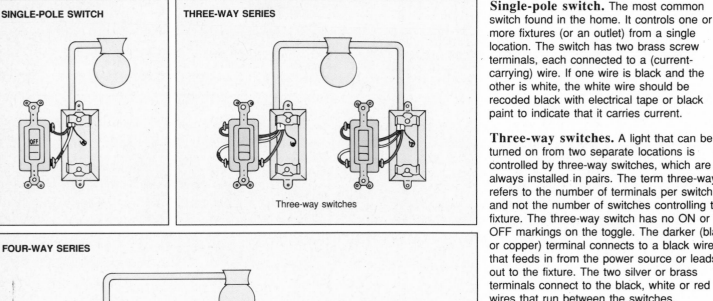

SINGLE-POLE SWITCH

THREE-WAY SERIES

Three-way switches

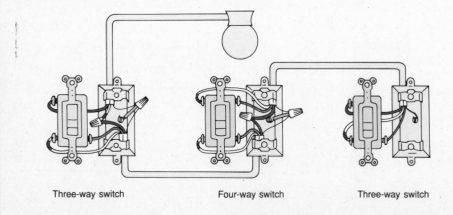

FOUR-WAY SERIES

Three-way switch   Four-way switch   Three-way switch

Single-pole switch. The most common switch found in the home. It controls one or more fixtures (or an outlet) from a single location. The switch has two brass screw terminals, each connected to a (current-carrying) wire. If one wire is black and the other is white, the white wire should be recoded black with electrical tape or black paint to indicate that it carries current.

Three-way switches. A light that can be turned on from two separate locations is controlled by three-way switches, which are always installed in pairs. The term three-way refers to the number of terminals per switch and not the number of switches controlling the fixture. The three-way switch has no ON or OFF markings on the toggle. The darker (black or copper) terminal connects to a black wire that feeds in from the power source or leads out to the fixture. The two silver or brass terminals connect to the black, white or red wires that run between the switches.

Four-way switches. Far less common than the three-way switch, the four-way switch is likely to be found near stairs or along hallways. It is always installed between a pair of three-way switches. The combination of three-way and four-way switches allows someone to operate a fixture from three or more locations. A four-way switch has four brass screw terminals and no ON or OFF markings on the toggle.

TROUBLESHOOTING GUIDE

Most switch problems can be solved by tightening the connections or replacing the switch. At right is a quick-reference guide to various switches in this chapter.

TYPE OF SWITCH	TIGHTEN CONNECTIONS	TEST AND REPLACE SWITCH
Single-pole switch (p. 74)	□○	□○
Dimmer switch (p. 79)	□○	■○ (Cannot be tested)
Time-delay switch (p. 80)	□○	□○
Time-clock switch (p. 81)	□○	■○
Ganged switch (p. 82)	□○	■○
Combination switch (p. 83)	□○	■○
Switch-outlet (switch and outlet independent) (p. 84)	□○	■○
Switch-outlet (switch controls outlet) (p. 85)	□○	■○
Pilot-light switch (p. 85)	□○	■○

SYMPTOM	POSSIBLE CAUSE	PROCEDURE
ALL SWITCH TYPES		
Fixture does not light or lights intermittently	Light bulb burned out	Replace light bulb
	Fuse blown or circuit breaker tripped	Replace fuse or reset circuit breaker (p. 18) □○
	Wires in box damaged	Repair wires (p. 104)
	Switch connections loose	Tighten connections (above)
	Switch faulty	Test and replace switch (above)
	Fixture faulty	Service fixture (p. 43)
Fuse blows or circuit breaker trips when switch flipped on	Too many fixtures or appliances on circuit	Move an appliance to another circuit
	Short circuit in wiring system	Check circuit (p. 104)
Sparks or crackling noise or light flickers when switch on	Switch connections loose	Tighten connections (above)
	Switch faulty	Test and replace switch (above)
ALL TOGGLE SWITCHES		
Toggle does not stay in ON position	Switch faulty	Test and replace switch (above)
THREE-WAY SWITCHES		
Both switches must be flipped to turn on fixture	Loose connection at one of the switches	Tighten connections of one switch at a time (p. .76) □○
	One of the switches faulty	Test switches one at a time and replace if faulty (p. 76) ■○
FOUR-WAY SWITCH SERIES		
Fixture does not light	Three-way switch connection loose	Tighten connections of one switch at a time (p. 76) □○
	One of the three-way switches faulty	Test switches one at a time and replace if faulty (p. 76) ■○
	Four-way switch connection loose	Tighten connections (p. 78) ■○
	Four-way switch faulty	Test and replace switch (p. 78) ■○
DIMMER SWITCH		
Light goes on but will not dim	Switch faulty	Replace switch (p. 79) ■○
TIME-DELAY SWITCH AND TIME-CLOCK SWITCH		
Fixture does not light at preset times	Timing mechanism worn	Test and replace switch (time delay, p. 80 □○; time clock, p. 81 ■○)
PILOT-LIGHT SWITCH		
Fixture lights but pilot light does not	Pilot light burned out	Replace switch (p. 85) ■○

DEGREE OF DIFFICULTY: □ **Easy** ■ **Moderate** ■ **Complex**
ESTIMATED TIME: ○ **Less than 1 hour** ◒ **1 to 3 hours** ● **Over 3 hours**

READING A SWITCH

The mounting strap is the place to find switch ratings and tester listing marks. The symbols UL and UND.LAB.INC.LIST confirm that the switch meets the Underwriters Laboratories safety standards. (A CSA monogram is the Canadian Standards Association stamp.) Maximum amperage and voltage ratings are also stamped on the mounting strap. For example, 15A-120V indicates that the switch can be used for up to 15 amperes of current at 120 volts. 15A-120VAC, or AC ONLY stamped elsewhere on the switch, indicates that it can only be used in alternating current systems, now almost universal in homes.

The two brass terminals of the single-pole switch are normally found on the side of the housing. The most common configuration of side-wired switches has an upper and lower terminal on the same side.

Some switches have a grounding terminal on the mounting strap at the top or bottom. The terminal is hexagonal, often colored green or marked GR. These switches are especially recommended for use in plastic boxes that do not contain a grounding bar *(page 73)*.

Wire specifications appear on the mounting strap and on the back of the switch. Switches rated at 30 amperes and above and with CU/AL markings accept copper, copper-clad aluminum, or aluminum wiring. A 15- or 20-ampere switch with a CO/ALR stamp can also be used with copper, copper-clad aluminum or aluminum wire. Older switches without these markings are for use with copper wire only. Prescribed wire size, usually ranging from No. 10 to No. 14, normally appears on the back; most circuits with light fixtures use No. 12 or No. 14 wire. On the back of some switches are push-in terminals and an accompanying wire strip gauge. When the switch has both push-in and screw terminals, hook the wires around the screw terminal, to provide a more secure connection.

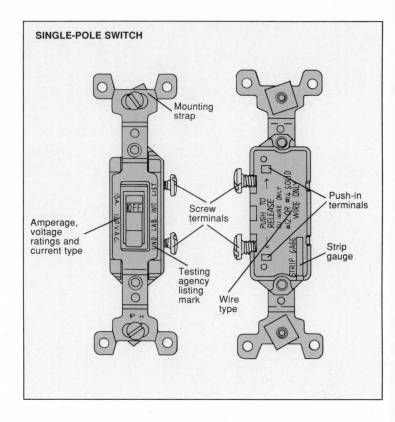

SINGLE-POLE SWITCH

Mounting strap

Amperage, voltage ratings and current type

Screw terminals

Push-in terminals

Strip gauge

Testing agency listing mark

Wire type

SWITCHES FOR SPECIAL USES

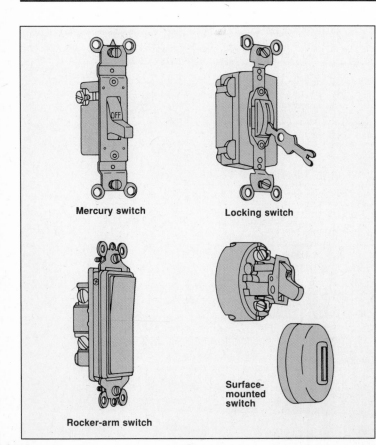

Mercury switch

Locking switch

Rocker-arm switch

Surface-mounted switch

Mercury switch. Because this switch has few moving parts, it has a long operating life. When the toggle is moved to the ON position, a pool of mercury bridges two contacts inside a sealed drum. Mercury switches may be single-pole or three-way. The word TOP is marked on the mounting strap to ensure correct installation.

Locking switch. This single-pole switch is frequently found in workshops or on other circuits that serve power tools. A special key, inserted in the slot on the face, operates the switch. To meet UL standards, the switch must be designed so that small metal objects inserted into the keyhole will not contact any electrical parts.

Rocker-arm switch. The large toggle of this switch makes it useful in areas such as laundry rooms, where the operator's hands may be full and an elbow is needed to flip a light on and off. Rocker-arm switches cost a little more than standard switches and are available in single-pole, three-way, four-way and lighted models.

Surface-mounted switch. This switch may be found in the garage or furnace room of an older house. Wires enter the side of the switch housing and attach to two screw terminals inside. The cover is removed by inserting a screwdriver in one of the slots on the side and gently turning to pry off the cover.

GROUNDING SWITCHES AND SWITCH BOXES

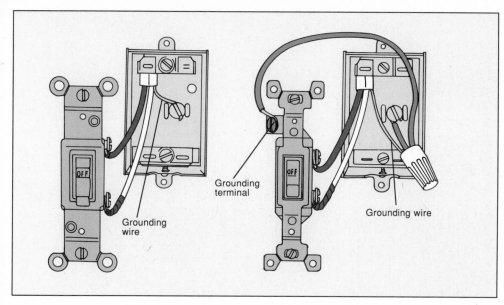

Grounding wire

Grounding terminal

Grounding wire

Grounding in metal boxes (one cable).
Grounding is an important safety precaution *(page 24)*. Replace a defective switch with one that has its own grounding terminal so that the switch can be grounded directly. If the switch does not have a grounding terminal, a grounded metal box provides a safe and reliable ground to the switch through its mounting screws.

When the switch does not have a grounding terminal, secure the bare copper grounding wire from the cable to the back of the metal box *(far left)*.

If the switch has a grounding terminal, fashion a pigtail *(page 141)* to ground the switch *(near left)*. Attach a jumper wire to the grounding terminal, and another to a screw at the back of the box. Twist the ends of these jumpers together with the bare wire from the cable and screw on a wire cap.

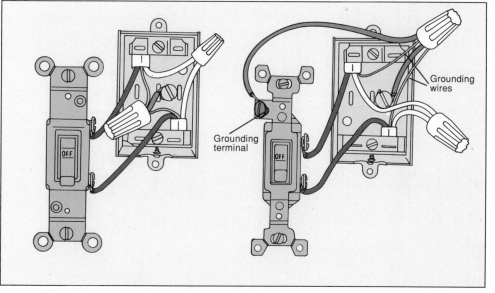

Grounding wires

Grounding terminal

Grounding in metal boxes (two cables).
If the switch does not have a grounding terminal, fashion a pigtail connection from the three grounding wires *(far left)*. Attach a jumper wire to the screw terminal on the back of the box, twist the other end of the jumper together with the bare grounding wires from each cable and screw on a wire cap. In Canada, code allows both of the grounding wires entering the box to be hooked around a screw on the back of the metal box.

When there is a grounding terminal on the switch, attach the jumper wire to the terminal and connect the jumper to the other grounding wires under a wire cap *(near left)*.

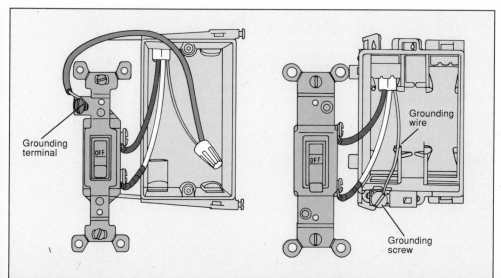

Grounding wire

Grounding terminal

Grounding screw

Grounding in plastic boxes. Where the switch has a grounding terminal and the box does not, attach a jumper wire to the switch grounding terminal and connect it to the bare grounding wire from the cable with a wire cap *(far left)*.

Some plastic boxes have a grounding screw on the metal bar. When there is no grounding terminal on the switch, ground the box by attaching the grounding wire to the metal bar *(near left)*. If the switch has a grounding terminal, make a pigtail as shown in the previous step to connect both the switch and box to the grounding wire from the cable. Where two cables enter the box, include the grounding wire from both cables in the pigtail.

SERVICING SINGLE-POLE SWITCHES

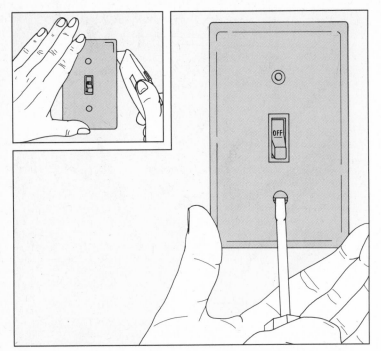

1 **Removing the cover plate.** Turn off power to the switch by removing the fuse or tripping the circuit breaker *(page 18)*. Remove the two screws on the cover plate *(above)* and lift it away from the wall. If the cover plate is stuck to the wall with paint or plaster, use a sharp knife to cut neatly around its edge *(inset)*. Tape the screws to the cover plate to avoid losing them.

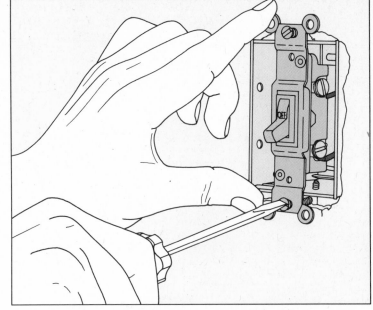

2 **Freeing the switch from the box.** Release the switch from the box by loosening the mounting screws located at the top and bottom of the mounting strap *(above)*. Grasp the mounting strap and pull the switch out of the box to expose the wires. Do not touch any wires or terminals until you have checked for voltage *(step 4)*.

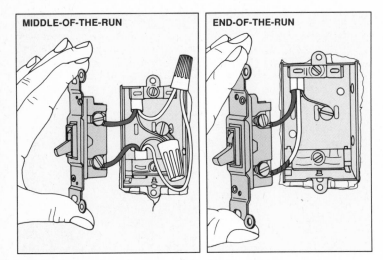

MIDDLE-OF-THE-RUN

END-OF-THE-RUN

3 **Identifying the wiring.** Check the number of cables entering the box. When the switch is middle-of-the-run, at least two cables enter the box *(above, left)*. Each cable has one black wire, one white wire, and a copper grounding wire. The black wires are hot (they carry current) and are connected to the switch terminals. The white wires are neutral and are joined inside the box with a wire cap. When the switch is end-of-the-run *(above, right)*, only one cable enters the box. Both the black and white wires connected to the switch terminals are hot. The white wire should be recoded black (hot) with electrical tape or black paint.

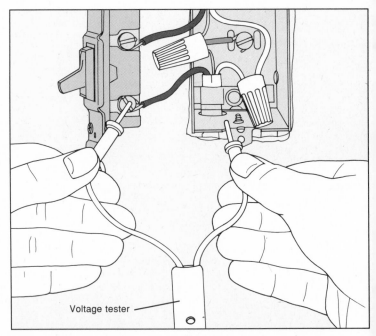

Voltage tester

4 **Testing for voltage.** Use a voltage tester to confirm that the power is off by touching one of the tester probes to the grounded metal box and the other probe to the terminal where a black wire is connected *(above)*. For plastic boxes, touch one probe to the switch terminal with the black wire and the other probe to the green grounding terminal on the switch (or to the metal grounding bar on the box or bare grounding wire). Repeat the procedure for the other switch terminal. The tester should not glow. If it does, return to the service panel and turn off power to the correct circuit.

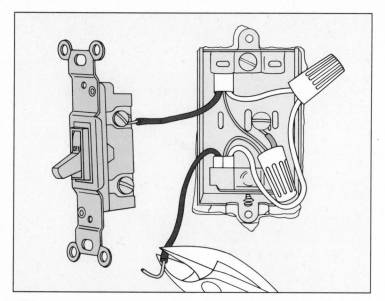

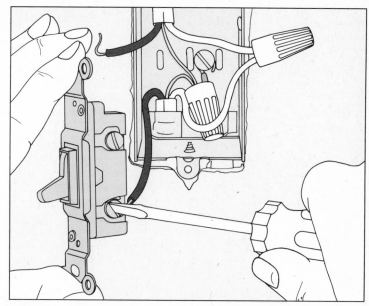

5 **Working on the connections.** Begin by checking the connections at the terminals. If a wire is loose, rehook it around the terminal and tighten the screw. If the connections appear dirty, detach the wires and clean the terminals and wire ends with fine sandpaper. Or clip the wires, as shown, and re-strip the insulation to expose clean wire *(page 141)*. With long-nose pliers, form the wire ends into a hook and secure each wire back under its terminal screw. Then screw the switch back into the box, turn on the power and flip on the switch to see if there is power to the fixture. If the switch works, put on the cover plate. If it doesn't work, turn off the power, take the switch out of the box and test it *(next step)*.

6 **Removing the switch.** Loosen the terminal screws and detach the wires, as shown. When there is a single cable in the box, mark the white wire with electrical tape or black paint, if not already marked. If there is a green grounding terminal on the switch, loosen the terminal screw and remove the grounding wire. If the wires are attached to push-in terminals on the back of the switch *(page 72)*, insert a small screwdriver blade into the release slot below each terminal and pull the wires free.

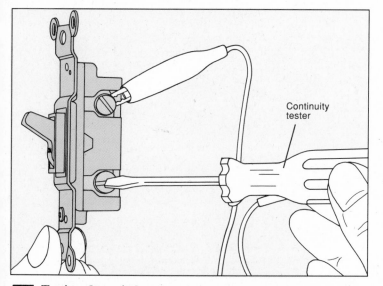

Continuity tester

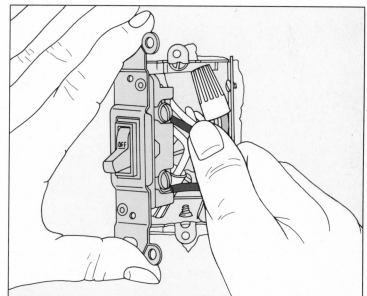

7 **Testing the switch.** Set the switch in the ON position. (Some switches have ON and OFF marked on the toggle; others have only a small, round bump to indicate the ON position.) Place the alligator clip of a continuity tester on one of the terminals and touch the tester probe to the opposite terminal, as shown. Then set the switch in the OFF position and repeat the procedure. On a good switch, the continuity tester will light when the switch is on, but not when the switch is off. If the switch passes the test, reconnect it, screw it back into the box, put on the cover plate and turn on the power. If the switch is defective, replace it *(next step)*.

8 **Installing the new switch.** Loosen the terminal screws of the new switch. Hold the switch so that the toggle points down when it is off. Hook the wires around the correct terminals: Where two cables enter the box, connect one black wire to each terminal; where one cable enters the box, connect the black wire to one terminal and the white wire to the other. Tighten the connections firmly. If there is a green grounding terminal on the switch, ground it *(page 73)*. Set the switch into the box, as shown, carefully folding the wires to make them fit. Screw the mounting strap back onto the box, making sure that the switch is straight. Put on the cover plate and turn on the power.

SERVICING THREE-WAY SWITCHES

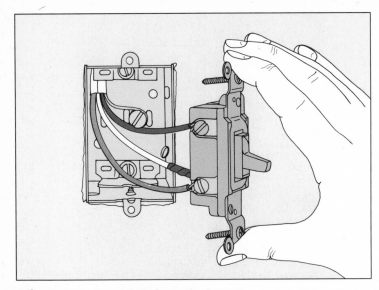

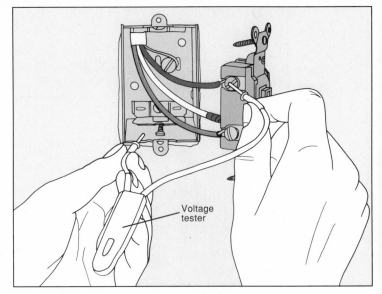

1 **Freeing the switch from the box.** Three-way switches are always installed in pairs *(page 70)*. To correct a problem in a three-way series, follow the procedure shown here for one of the switches; if it is not faulty, repeat the procedure for the other. Start by turning off power to the switch at the service panel *(page 18)*. Remove the screws on the cover plate and lift it away from the wall. Loosen the screws on the mounting strap and pull the switch out of the box to expose the wires, as shown. Whether one or two cables enter the box, there are always three wires connected to the switch. The wire that feeds in from the power source or leads to the fixture is attached to the darker screw terminal. The two wires that run between the switches are connected to the silver or brass terminals.

2 **Testing for voltage.** Use a voltage tester to confirm that the power is off by touching one of the tester probes to the grounded metal box and the other probe to one of the three switch terminals *(above)*. For plastic boxes, touch one probe to a switch terminal and the other probe to the green grounding terminal on the switch (or the metal grounding bar on the box or bare grounding wire). Repeat the procedure for the other two switch terminals. The tester should not glow. If it does, return to the service panel and find the correct fuse or circuit breaker and turn off power to the switch.

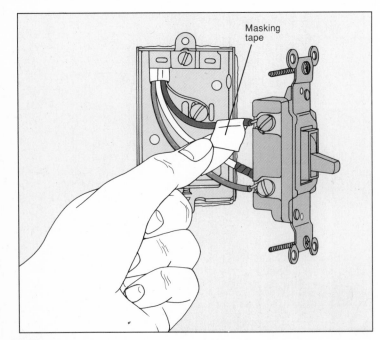

Masking tape

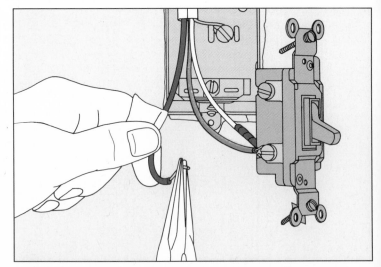

3 **Marking the wires.** All three wires connected to the switch are hot (they carry current). A black wire is attached to the darker (copper or black) terminal. To help you when reconnecting the switch, mark this wire with a piece of masking tape, as shown. The other two wires are connected to the silver or brass terminals. One is red, the other is black (or white recoded black). If the white is not recoded black to indicate that it carries current, mark the end with electrical tape or black paint.

4 **Working on the connections.** Check the connections at the terminals. If a wire is loose, rehook it around the terminal and tighten the screw. If the connections appear dirty, clean the terminals and wire ends with fine sandpaper. It may be necessary to snip back the wire ends and re-strip the insulation *(page 141)*. Then recurl the wire ends *(above)* and hook each wire back around its terminal. If there are two cables in the box, check the connections under the wire caps. Screw the switch back into the box, turn on the power and flip on the switch to see if there is power to the fixture. If the switch works, put on the cover plate. If it doesn't, turn off the power again and take the switch out of the box *(next step)*.

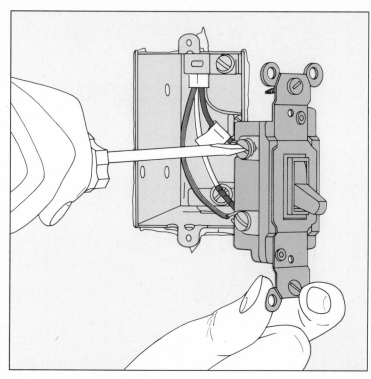

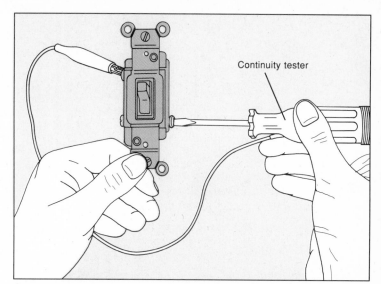

5 **Removing the old switch.** Loosen the screw terminals *(above)*. Disconnect the tagged wire from the darker terminal. Disconnect the other two wires from their terminals and remove the switch.

6 **Testing the switch.** Place the alligator clip of a continuity tester on the darker (black or copper) terminal. Touch the tester probe to one of the other terminals, as shown. Flip the switch on, then off. The tester should glow in one position and not in the other. Set the toggle to the position in which the tester glows and touch the tester probe to the opposite terminal. The tester should not glow with the toggle in this position but should glow when the toggle is moved to the other position. If the switch fails any of these tests, replace it *(next step)*. If the switch passes the tests, reconnect it, put it back in the box and put on the cover plate. Then check the other three-way switch.

Continuity tester

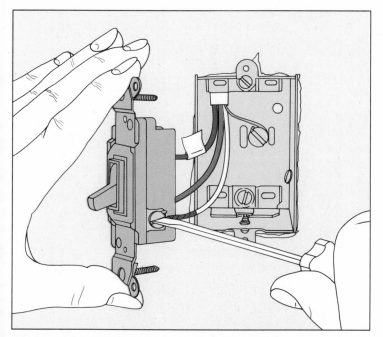

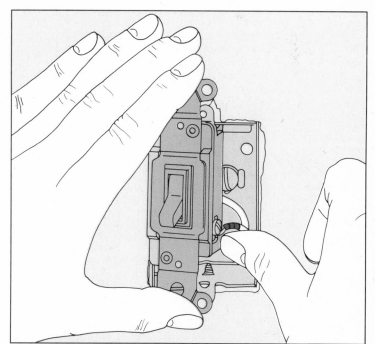

7 **Connecting the new switch.** Hook the tagged wire around the black or copper terminal of the new switch and tighten the screw. Connect the other two wires to their terminals in the same way *(above)*. Either wire can be connected to either terminal. If there is a green grounding terminal on the switch, ground it *(page 73)*.

8 **Setting the switch into the box.** Grasp the switch by the mounting strap and push it into the box, as shown, carefully folding the wires to make them fit. Screw the mounting strap back onto the box, making sure the switch is straight. Put on the cover plate and turn on the power.

SERVICING FOUR-WAY SWITCHES

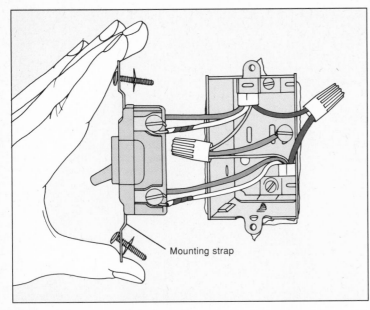

Mounting strap

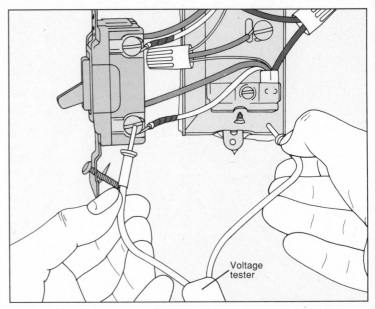

Voltage tester

1 **Freeing the switch from the box.** Turn off power to the switch at the service panel *(page 18)*. Remove the cover plates of all switches that operate the fixture. Identify the three-way and four-way switches: Three-ways have three terminals, four-ways have four terminals. Begin by servicing the three-way switches *(page 76)*; if none of these is faulty, test the four-way switch in the series. Loosen the screws on the mounting strap and pull the switch from the box to expose the wires. There are two cables entering the box; all four wires connected to the switch are hot (they carry current). Two wires connected to one side of the switch are red; two attached to the other side are either black or white recoded black *(above)*.

2 **Testing for voltage and checking the connections.** Use a voltage tester to confirm that the power is off by touching one of the tester probes to the grounded metal box and the other probe to one of the four screw terminals *(above)*. For plastic boxes, touch one probe to the switch terminal and the other probe to the green grounding terminal on the switch (or to the metal grounding bar on the box or bare grounding wire). Repeat the procedure for all four terminals. The tester should not glow. If it does, return to the service panel and turn off power to the correct circuit. Next check for loose or dirty connections. Clean the wires and screw terminals, and re-strip the wires if necessary *(page 141)*. Tighten the terminal screws and wire caps. Screw the switch back into the box and turn on the power. If the switch works, put on the cover plate. If it doesn't, turn off the power, take the switch out of the box and go to step 3.

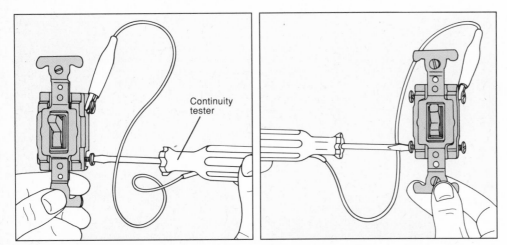

Continuity tester

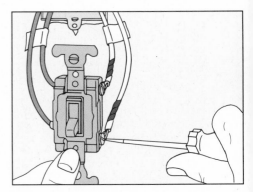

3 **Testing the switch.** To help you when reconnecting the switch, use masking tape to tag the wires connected to the two upper terminals. Then disconnect the switch. Place the alligator clip of a continuity tester on one of the upper terminals. Touch the tester probe to the lower terminal on the same side *(above, left)*. If the tester fails to light, set the switch toggle to the other position. The tester should light. Next, touch the tester probe to the lower terminal on the opposite side *(above, right)*, and move the toggle to the other position. The tester should light. Remove the alligator clip from the terminal, place it on the upper terminal on the opposite side, and repeat the procedure. The tester should not light. If the switch fails any of these tests, install a replacement *(step 4)*. If the switch is good, reconnect it, put it back in the box and put on the cover plate. Then check any other four-way switches in the series.

4 **Installing the new switch.** Hook the tagged red wire around an upper terminal. Attach the tagged white wire to the upper terminal on the opposite side. Connect the other red and white wires as shown. Screw the switch back into the box, making sure that it is straight. Put on the cover plate and turn on the power.

SERVICING DIMMER SWITCHES

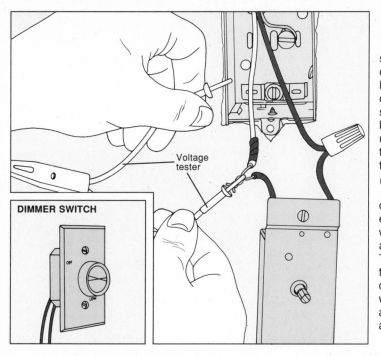

Voltage tester

DIMMER SWITCH

OFF

LOW

1 **Removing the old dimmer switch.** A dimmer switch has a control knob or toggle that not only turns a fixture on and off but also adjusts the lighting level from low to high. To service this switch, begin by turning off power at the service panel *(page 18)*. Pull off the control knob, then remove the cover plate. Unscrew the switch housing and ease it from the box. If the dimmer switch has two leads or screw terminals *(inset)*, it is single-pole; if it has three leads or screw terminals, it is three-way. A single-pole dimmer switch may have one or two cables entering the box. For a switch with leads, unscrew the wire caps to expose the connections (but do not touch the wire ends). For a switch with screw terminals, test to confirm that the power is off and disconnect it as you would a regular single-pole *(page 74)* or three-way switch *(page 76)*.

Use a voltage tester to confirm that the power is off by touching one of the tester probes to the grounded metal box and the other to an exposed wire connection *(left)*. For plastic boxes, touch one probe to a wire connection and the other to the grounding terminal on the box or a bare grounding wire. Repeat this procedure for the other connection. The tester should not glow. If it does, return to the service panel and turn off power to the correct circuit. Detach the wires and clean them or clip them back *(page 141)*. Then twist them together and screw the wire caps back on securely. Put the dimmer switch back in the box and turn on the power. If the switch does not work, turn off the power again, remove the switch and install a new one.

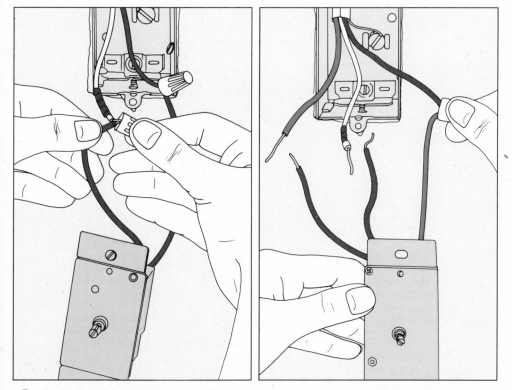

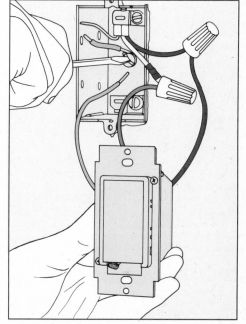

2 **Installing the new dimmer switch.** Buy a replacement switch with the same wattage and number of leads or terminals as the old one. To install a single-pole, touch-sensitive dimmer switch, follow the step at right. To connect a regular single-pole dimmer switch, attach each lead to a wire and secure the connections with wire caps *(above, left)*. It doesn't matter which lead is attached to which wire.

For a three-way dimmer switch, twist the red lead from the switch to the black wire and screw on a wire cap *(above, right)*. The other two switch leads are the same color (both red or both black). Connect each switch lead to a wire and twist a wire cap onto each. (It doesn't matter which lead is attached to which wire.) Screw the switch housing into the box, screw on the cover plate and push the control knob in place. Press the knob to turn on the switch, then turn on the power.

Installing a touch-sensitive dimmer switch. These switches have a touch-sensitive panel in place of a control knob, and three leads: black, red and green. Attach the black lead to the black wire from the box, and the red lead to the white wire from the box. The green lead is the grounding wire. Connect a jumper wire to the grounding screw at the back of the box, as shown, then twist the green lead together with the jumper wire and the bare grounding wire entering the box. Screw on a wire cap. For a plastic box, attach the green lead to the bare grounding wire from the cable.

REPLACING A TIME-DELAY SWITCH

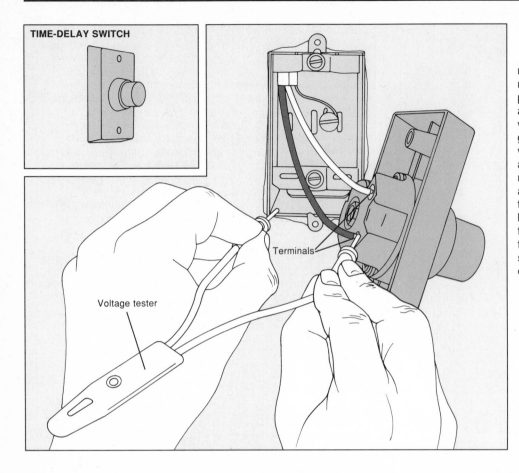

TIME-DELAY SWITCH

Terminals

Voltage tester

1 **Testing for voltage.** The time-delay switch *(inset)* is usually found in basements and garages, where the operator needs to light an area only long enough to reach another lighted area. When the knob is pressed, it turns on a fixture, then flicks it off automatically after up to 45 seconds. To service a time-delay switch, which is always single-pole, begin by turning off power at the service panel *(page 18)*. Remove the cover plate and pull the switch from the wall, turning it to reveal the push-in terminal holes. Use a voltage tester to confirm that the power is off by touching one probe to the grounded metal box, the other to a terminal, as shown. Repeat the procedure for the other terminal. The tester should not glow. If it does, return to the service panel and turn off power to the correct circuit.

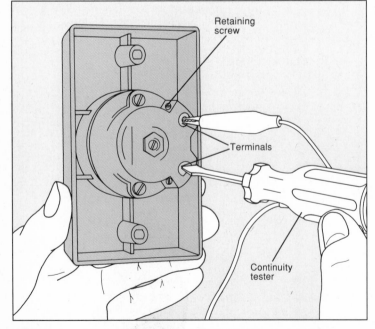

Retaining screw

Terminals

Continuity tester

2 **Removing and testing the switch.** Tighten the retaining screws to secure the wires to their push-in terminals, then return the switch to the box, turn on the power and press the knob to see if it works. If it doesn't, turn off the power and detach the wires from their terminals. To test the switch, press the knob, insert the alligator clip of the tester into a terminal hole and touch the tester probe to the other terminal, as shown. The continuity tester should light while the knob is depressed. If it doesn't, replace the switch.

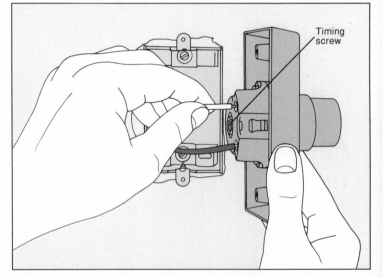

Timing screw

3 **Installing the new time-delay switch.** Before wiring the new switch, adjust the timing screw with a screwdriver, turning clockwise for a longer delay and counterclockwise for a shorter one. Then push a wire into each of the terminal holes *(above)*, making sure no uninsulated wire is exposed. If necessary, clip the wire ends to ensure a proper fit. Tighten the retaining screws to secure the connections. Screw the switch into the box and turn on the power.

REPLACING A TIME-CLOCK SWITCH

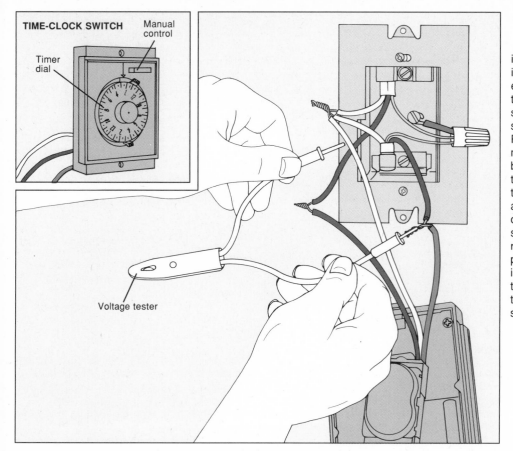

TIME-CLOCK SWITCH

Timer dial

Manual control

Voltage tester

1 Testing for voltage. A time-clock switch *(inset)* turns a fixture on and off automatically with delay periods ranging from several minutes to 24 hours. It is installed only where two or more cables enter the box. (The white wires are attached to the timer motor.) To replace a time-clock switch, begin by turning off power to the switch at the service panel *(page 18)*. Remove the screws securing the timer to its mounting plate and free the timer from the box. Twist off the wire caps, taking care not to touch the bare wire ends. To confirm that the power is off, touch one probe of a voltage tester to the mounting plate and the other to each of the wire connections, as shown. The tester should not glow. If it does, return to the service panel and turn off power to the correct circuit. Once the power is off, detach the wires and clean or clip them back *(page 141)*. Then twist them together and screw the wire caps back on securely.

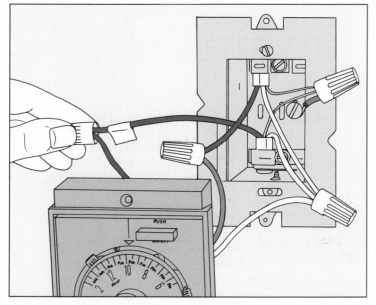

2 Removing and testing the switch. After tightening the connections, screw the time-clock switch back into the box and turn on the power, then press the manual control. If the switch still does not work, turn off the power and pull the switch from the box. Use tape to tag the black wire that is joined to the timer's black lead, then undo the wire connections *(above)*. Attach the alligator clip of a continuity tester to the black lead and touch the tester probe to the red lead, then press the manual control to turn on the switch. The tester should light when the switch is turned on. If it does not light, unscrew the switch mounting plate and replace the switch *(next step)*.

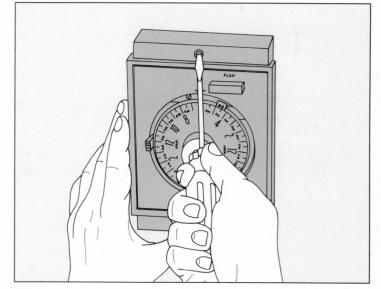

3 Installing the new time-clock switch. Replace the time-clock switch with a similar model, or install a new digital timer-dimmer. If you are installing the same model, begin by securing the new mounting plate to the box. Twist the black switch lead together with the tagged black wire and secure the connection with a wire cap. Join the red switch lead to the untagged black wire in the same manner. Twist the white switch lead together with the pigtail of white wires entering the box, and screw on a wire cap. Screw the timer to the mounting plate *(above)*, turn on the power and press the manual control to test.

SERVICING GANGED SWITCHES

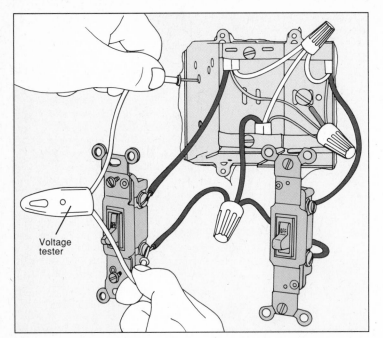

Voltage
tester

1 **Testing for voltage.** Turn off power to the switch at the main service panel *(page 18)*. Remove the screws in the cover plate and lift it away from the wall. Loosen the screws on the mounting straps and pull both switches from the box to expose the wires. If a black and white wire are attached to each switch, the switches are on separate circuits. Go to step 3 to service the faulty switch. When one black wire is connected to both switches by means of jumper wires, as in the illustration above, the switches are on the same circuit. Use a voltage tester to confirm that the power is off by touching one probe to the grounded metal box and the other to one of the terminals, as shown, turning the switch on and then off. Repeat the procedure for all terminals on both switches. The tester should not glow. If it does, return to the service panel and turn off the power to the correct circuit.

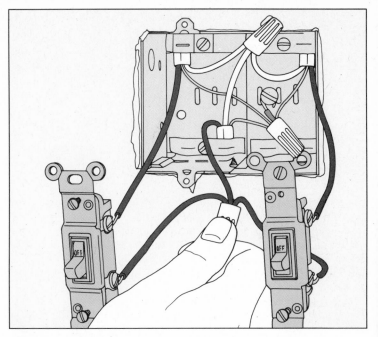

2 **Checking the connections.** Unscrew the wire caps to examine the connections. Clean the wires with fine sandpaper, twist them back together and replace the caps, as shown. Then clean and retighten the terminal connections. Screw the switches back into the box, turn on the power and flip on the switch. If the fixture does not light, turn off the power and remove the switch. Test and replace it as you would a regular single-pole switch *(page 74)*.

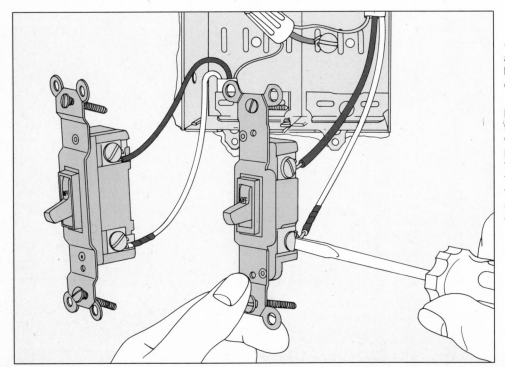

3 **Servicing a ganged switch on a separate circuit.** When there are two switches in a box and they are on separate circuits, make sure that power is off to both circuits before beginning any repairs on either switch.

Use a voltage tester to confirm that the power is off, following the procedure for single-pole switches *(page 74)*. Once you have confirmed the power is off to both switches, push the good switch aside and work only on the faulty switch *(left)*. Service the connections and test and replace the switch as you would a single-pole switch *(page 74)*. Then screw both switches back in the box and turn on the power to both circuits.

REPLACING A COMBINATION SWITCH

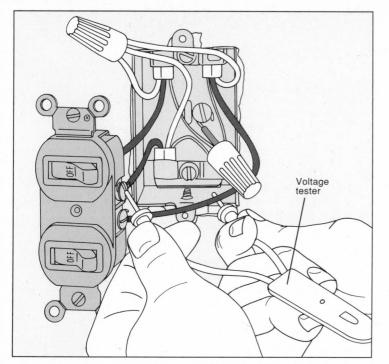

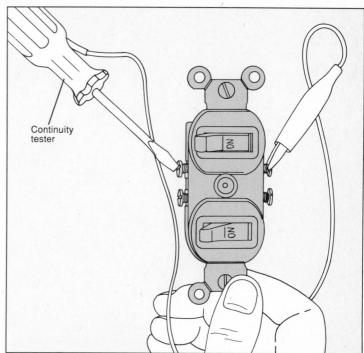

1 **Testing for voltage and servicing the connections.** A combination switch has two switch toggles in a single housing. To repair it, first turn off the power at the service panel *(page 18)*. Remove the cover plate, loosen the mounting screws and free the switch from the box. Use a voltage tester to confirm that power is off by touching one probe to the box and the other to each screw terminal in turn *(above)*. The tester should not glow. If it does, return to the service panel and turn off power to the correct circuit. Clean and tighten the connections, then return the switch to the box, turn on the power and flip on both toggles. If either does not work, turn off the power and remove the switch from the box. To help you when reconnecting the switch, use masking tape to tag the wire attached to the terminal with the connecting tab. (The connecting tab on the side of the switch ensures the flow of power to both terminals, and thus to both switch toggles.)

2 **Testing the switch.** Disconnect the switch. With the upper toggle in the OFF position, place the alligator clip of a continuity tester on one upper terminal and touch the tester probe to the other upper terminal. Then flip on the toggle and repeat the test, as shown. The tester should glow when the toggle is on and not glow when it is off. Repeat the procedure for the lower toggle. If the switch tests faulty, replace it.

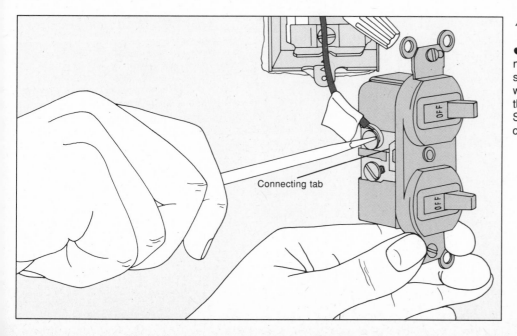

3 **Installing the new switch.** Hook the incoming (tagged) black wire around the terminal on the side with the connecting tab and tighten the connection, as shown. Then connect the other two black wires to the upper and lower terminals on the opposite side of the switch and tighten. Screw the switch into the box, put on the cover plate and turn on the power.

REPLACING A SWITCH-OUTLET (Switch and outlet are independent)

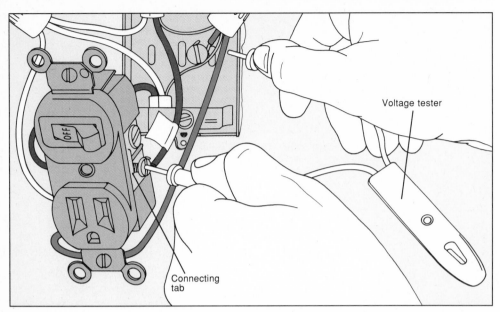

Voltage tester

Connecting tab

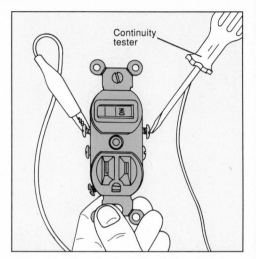

Continuity tester

1 **Testing for voltage and servicing the connections.** Depending on their wiring, the switch and outlet can operate independently *(above)*, or the switch can control the outlet *(page 85)*. First turn off power to the switch-outlet at the service panel *(page 18)*. Unscrew the cover plate, loosen the mounting strap and pull the switch-outlet free from the box. Use a voltage tester to confirm that the power is off by touching one probe to the box and the other to one screw terminal, as shown, then to the other. The tester should not glow. If it does, return to the service panel and turn off the power to the correct circuit. Clean and tighten the wire connections, then screw the switch back into the box, turn on the power and test the switch or outlet. If the problem persists, turn off the power again and tag the black wire connected to the terminal on the side with the connecting tab. (The connecting tab ensures the flow of power to both terminals, and thus to both the switch and outlet.)

2 **Testing the switch-outlet.** Disconnect the switch-outlet. With the switch toggle in the OFF position, place the alligator clip of a continuity tester on an upper terminal and touch the tester probe to the other upper terminal. Then set the toggle to the ON position and repeat the test, as shown. On a good switch-outlet, the tester should glow when the switch is in the ON position, but not when it is in the OFF position. If the switch fails either test, replace it. To connect a switch-outlet where the outlet is always hot, go to step 3. To install one where the switch controls the outlet, follow the instructions on page 85.

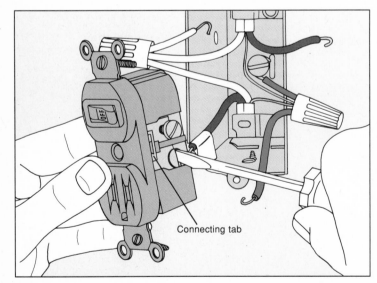

Connecting tab

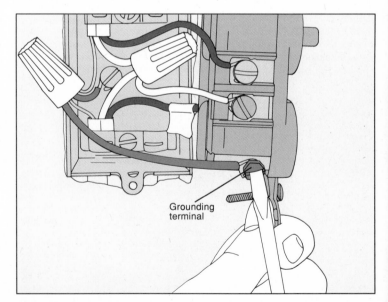

Grounding terminal

3 **Connecting the switch-outlet.** Hook the tagged black wire around a brass terminal on the side of the switch with the connecting tab and tighten, as shown. The second black wire leads to the fixture on the circuit. Wrap it around the upper terminal on the opposite side and tighten firmly. The switch portion of the switch-outlet is now hooked up. An outlet always needs a neutral (white) wire to complete the circuit. Attach the white jumper wire to the lower, silver terminal on the side with no connecting tab and tighten.

4 **Grounding the switch-outlet.** Locate the green grounding terminal on the lower part of the switch-outlet. Connect the grounding jumper to the grounding terminal on the switch-outlet and tighten, as shown. Screw the switch-outlet into the box and turn on the power. Plug a lamp into the outlet and turn it on. If the outlet does not work, the incoming and outgoing black wires may have been reversed. Turn off the power again and change the black wire connections. Then set the switch-outlet back in the box, turn on the power and test the outlet again.

CONNECTING A SWITCH-OUTLET (Switch controls the outlet)

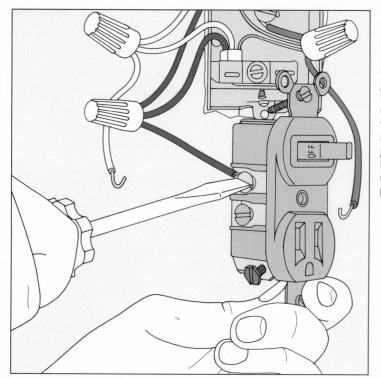

Installing a replacement switch.
Turn off the power, service the connections and perform a continuity test as described on page 84. To install a replacement where the switch controls the outlet, fasten the black jumper wire to the upper terminal on the side with no connecting tab, and tighten the connection, as shown. Hook the white jumper wire around the lower terminal on the same side and tighten. Then attach the grounding jumper wire to the grounding terminal on the lower part of the outlet and tighten. Screw the switch-outlet into the box and turn on the power.

REPLACING A PILOT-LIGHT SWITCH

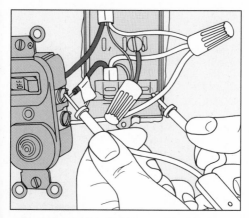

1 Testing for voltage and servicing the connections. The pilot-light switch indicates that a light is on in a part of the house where the fixture is not readily visible (such as a basement or attic). To service it, turn off power to the switch at the service panel *(page 18)*, and pull the switch from the box. Use a voltage tester to confirm that the power is off by touching one probe to the grounded metal box and the other to each terminal, in turn *(above)*. The tester should not glow. If it does, return to the service panel and turn off power to the correct circuit. If the pilot light is burned out, the switch must be replaced *(step 3)*.

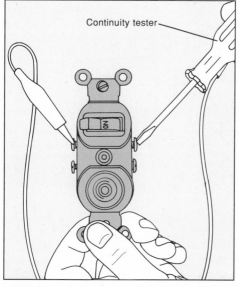

Continuity tester

2 Testing the switch. Clean and tighten the connections, then return the switch to the box, turn on the power and flip on the switch. If the problem persists, turn off the power again and tag the wire attached to the upper terminal on the side with no connecting tab. Then disconnect the switch and flip on the toggle. Using a continuity tester, place the alligator clip on one upper terminal and touch the tester probe to the opposite terminal, as shown. If the tester doesn't light, replace the switch *(step 3)*.

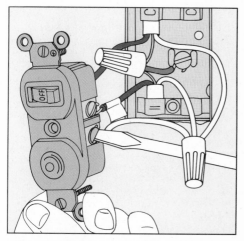

3 Installing the new switch. Attach the tagged black wire to the upper terminal on the side of the switch that has no connecting tab. Hook the other black wire around the upper terminal on the other side. Hook the jumper wire from the white, neutral pigtail around the lower terminal below the tagged black wire, as shown. Tighten the connections. Return the switch to the box, turn on the power and flip the switch on and off. If the fixture and the pilot light work, put on the cover plate.

WALL OUTLETS

The common wall outlet is essentially inert. Current arrives at its hot side and is kept on hold until an appliance is plugged into it, completing the circuit. A standard 15-ampere, 120-volt, grounded duplex outlet has an upper and lower receptacle, each with three slots. It is a sturdy and long-lasting device; any problems can usually be traced to the lamp or appliance plugged into the outlet rather than the outlet itself. When it does wear out, or it no longer makes contact with the plug prongs, an outlet can be quickly and inexpensively replaced.

Once the cover plate is removed, you will find one of a number of wiring variations inside an outlet box. The number of cables is determined by the location of the outlet along the circuit. An outlet that is installed mid-circuit (middle-of-the-run) will have two or more cables in the box: one cable coming in from the power source and the other going out to the next box on the circuit. An outlet box located at the end of the circuit (end-of-the-run) has a single incoming cable. An outlet is also wired differently if it is controlled by a switch, or if it is supplied by a 240-volt split circuit *(page 95)*.

The trick to attaching an outlet is simply to connect the correct wire to the correct terminal. Black (sometimes red) wire connects to brass terminals and white wire connects to silver terminals. Upper and lower terminals on the same side of the outlet operate as a single unit and may be used interchangeably until the connecting tab is removed. Where the wiring does not follow this pattern exactly, the differences are described within each repair.

In addition to the standard 120-volt duplex outlet, there is a variety of outlets for special needs. The GFCI (ground-fault circuit interrupter) is an important safety feature required by code in bathrooms, garages and outdoor circuits. It monitors the flow of electricity at an outlet and shuts off all current in the event of a leak.

Heavy-duty, 240-volt outlets—which will accept only matching plugs—are found behind most major appliances. Both the outlet and its box must be grounded *(page 89)* to protect against shock in the event of a short circuit. A two-slot, 120-volt outlet will not admit a grounded (three-prong) plug and should be replaced with a grounded three-slot outlet. If there is no grounding wire in the box, replace a two-slot outlet with a GFCI outlet.

When working in an outlet box, proceed carefully, especially if it is supplied with 240-volt current. Work only in dry conditions. Do not touch any terminals or wire ends until you have turned off the power to the outlet and used a voltage tester to confirm that the power is off.

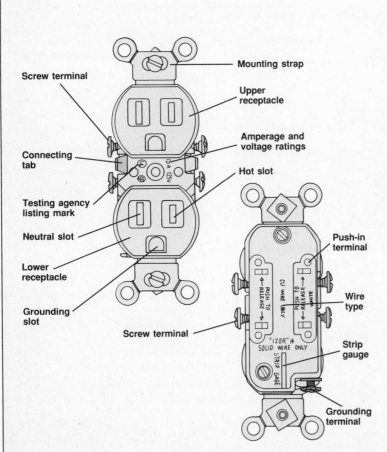

Reading an outlet. The standard grounded duplex outlet has an upper and lower receptacle. Each receptacle has a long (neutral) slot for the wide prong of a plug, a shorter (hot) slot for the narrow prong, and a U-shaped grounding slot.

Look on the front and back of the outlet to find ratings and tester stamps. The symbol UL, or UND.LAB.INC.LIST, confirms that the outlet meets the Underwriters Laboratories standardized tests. (CSA is the Canadian Standards Association monogram.) Voltage and amperage ratings, stamped on the front of the outlet, stipulate the maximum levels at which the outlet can be safely used. For example, 15A-125V indicates that the outlet is designed for up to 15 amperes of current at 125 volts.

Two brass terminals are found on one side of the outlet and two silver terminals on the other side. Sometimes WHITE is stamped on the back of the outlet to identify the side with the silver terminals. A connecting tab links the terminals on each side; when the tab between the brass terminals is snapped off, the upper and lower receptacles function independently. The grounding terminal of the outlet is hexagonal and colored green.

Wire specifications appear on the mounting strap or the back of the outlet. Outlets rated at 30 amperes and above and with a CU/AL marking take copper, copper-clad aluminum or aluminum wiring. A 15- or 20-ampere outlet with a CO/ALR stamp can be used with copper, copper-clad aluminum or aluminum wiring. Older, unmarked 15- or 20-ampere outlets take only copper or copper-clad; older, unmarked 30-ampere outlets are for use with copper wire only. Prescribed wire size, usually ranging from No. 10 to No. 14 gauges, normally appears on the back. On the back of some outlets there are also push-in terminals and an accompanying wire-strap gauge. When the outlet has both push-in and screw terminals, hook the wires around the screw terminals to provide a more secure connection.

TROUBLESHOOTING GUIDE

Most outlet problems can be solved by servicing the connections or replacing the outlet. At right is a quick-reference guide to various outlets in this chapter.

TYPE OF OUTLET	TIGHTEN CONNECTIONS	TEST AND REPLACE OUTLET
Standard duplex outlet (p. 90)	□○	■○
Switch-controlled outlet (switch controls upper and lower receptacles) (p. 92)	□○	■○
Switch-controlled outlet (switch controls one receptacle) (p. 93)	□○	■○
Split-circuit outlet (p. 95)	□○	■○
GFCI outlet (p. 96)	□○	■○
120/240-volt outlet (p. 100)	□○	■◕
240-volt outlet (p. 101)	□○	■◕

SYMPTOM	POSSIBLE CAUSE	PROCEDURE
ALL OUTLET TYPES		
Appliance in outlet does not work	Appliance faulty	Check appliance
	No power to outlet	Replace fuse or reset circuit breaker (p. 18) □○
	Outlet connections loose	Service connections (above)
	Outlet faulty	Replace outlet (above)
Fuse blows or breaker trips when appliance is plugged in	Circuit overloaded	Move appliance to another circuit
	Appliance faulty	Check appliance
	Short circuit in wiring system	Check circuit (p. 104)
Appliance plug falls out easily	Appliance plug faulty	Check appliance plug
	Outlet worn	Test outlet with reliable plug; if plug falls out, replace outlet (above)
Appliance runs intermittently or lamp flickers	Appliance or lamp faulty	Check appliance or lamp
	Outlet connections loose	Service connections (above)
Sparks or mild shock when plugging in appliance	Appliance is turned on	Turn off appliance before plugging it in
	Fingers contacting plug prongs	Hold plug by insulation
	Appliance faulty	Check appliance
	Outlet connections loose	Service connections (above)
SWITCH-CONTROLLED OUTLET		
Outlet does not deliver power	Wall switch faulty	Check wall switch (p. 70)
SPLIT-CIRCUIT DUPLEX OUTLET		
Only one receptacle in duplex works	Outlet connections loose	Service connections (p. 95) □○
GFCI OUTLET		
GFCI trips constantly	Appliance faulty	Check appliance
	Outlet connections loose	Service connections (p. 96) □○
	GFCI faulty	Replace outlet (p. 96) ■◕
	Faulty appliance or wiring in box elsewhere on same circuit	Check appliances on circuit Check wiring in boxes on circuit (p. 104)
TWO-SLOT OUTLET		
Outlet will not accept grounded (three-prong) plugs	Older outlet	Replace two-slot outlet with three-slot outlet (p. 98) ■◕ Install grounding-adapter plug (p. 99) □○

DEGREE OF DIFFICULTY: □ Easy ■ Moderate ■ Complex
ESTIMATED TIME: ○ Less than 1 hour ◕ 1 to 3 hours ● Over 3 hours

120-VOLT OUTLETS

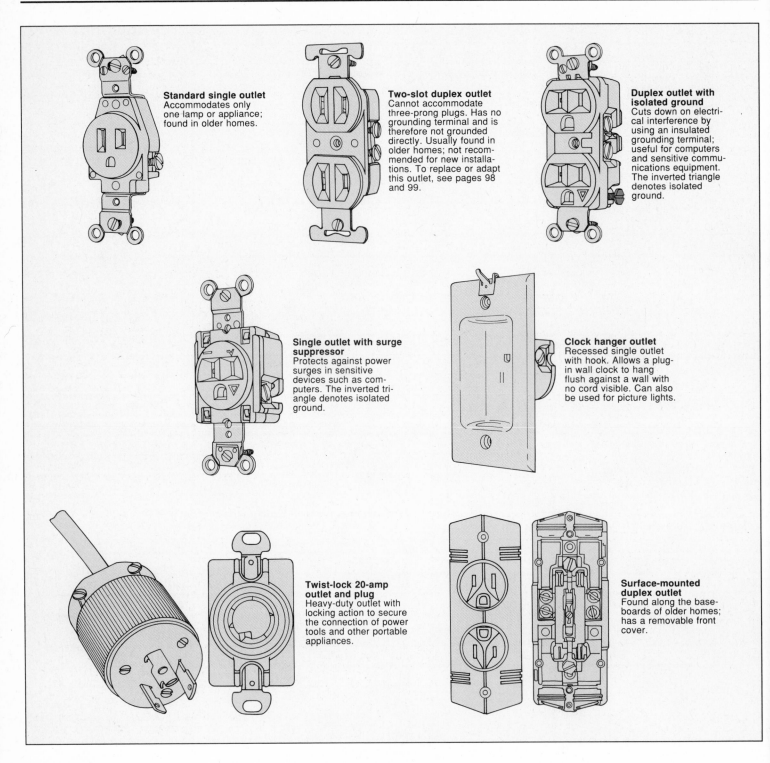

Standard single outlet
Accommodates only
one lamp or appliance;
found in older homes.

Two-slot duplex outlet
Cannot accommodate
three-prong plugs. Has no
grounding terminal and is
therefore not grounded
directly. Usually found in
older homes; not recom-
mended for new installa-
tions. To replace or adapt
this outlet, see pages 98
and 99.

**Duplex outlet with
isolated ground**
Cuts down on electri-
cal interference by
using an insulated
grounding terminal;
useful for computers
and sensitive commu-
nications equipment.
The inverted triangle
denotes isolated
ground.

**Single outlet with surge
suppressor**
Protects against power
surges in sensitive
devices such as com-
puters. The inverted tri-
angle denotes isolated
ground.

Clock hanger outlet
Recessed single outlet
with hook. Allows a plug-
in wall clock to hang
flush against a wall with
no cord visible. Can also
be used for picture lights.

**Twist-lock 20-amp
outlet and plug**
Heavy-duty outlet with
locking action to secure
the connection of power
tools and other portable
appliances.

**Surface-mounted
duplex outlet**
Found along the base-
boards of older homes;
has a removable front
cover.

ADAPTERS AND EXTENSIONS

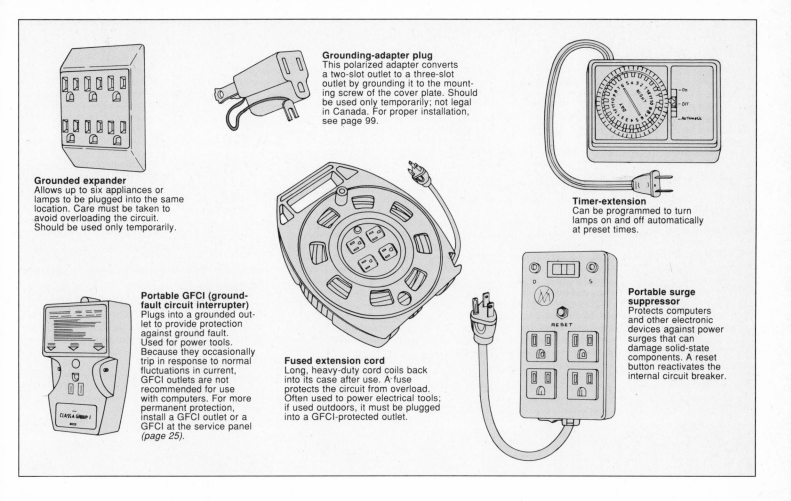

Grounded expander
Allows up to six appliances or lamps to be plugged into the same location. Care must be taken to avoid overloading the circuit. Should be used only temporarily.

Grounding-adapter plug
This polarized adapter converts a two-slot outlet to a three-slot outlet by grounding it to the mounting screw of the cover plate. Should be used only temporarily; not legal in Canada. For proper installation, see page 99.

Timer-extension
Can be programmed to turn lamps on and off automatically at preset times.

Portable GFCI (ground-fault circuit interrupter)
Plugs into a grounded outlet to provide protection against ground fault. Used for power tools. Because they occasionally trip in response to normal fluctuations in current, GFCI outlets are not recommended for use with computers. For more permanent protection, install a GFCI outlet or a GFCI at the service panel (page 25).

Fused extension cord
Long, heavy-duty cord coils back into its case after use. A fuse protects the circuit from overload. Often used to power electrical tools; if used outdoors, it must be plugged into a GFCI-protected outlet.

Portable surge suppressor
Protects computers and other electronic devices against power surges that can damage solid-state components. A reset button reactivates the internal circuit breaker.

GROUNDING IN OUTLET BOXES

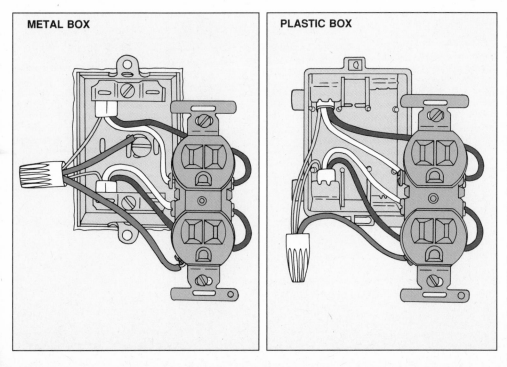

METAL BOX

PLASTIC BOX

Grounding in metal boxes. All three-slot outlets must be grounded (page 24). To ground an outlet in a metal box (far left), attach a jumper wire to the grounding screw at the back of the box and another to the grounding terminal on the outlet. Then twist the ends of these jumpers together with the bare wires entering the box and screw on a wire cap. (In Canada, code allows the bare grounding wire entering the box and the jumper on the outlet to be connected directly to the grounding screw at the back of the box.) Some armored cable has no grounding wire; to ground the outlet, attach a jumper to the grounding screw at the back of the box and the grounding terminal on the outlet.

Grounding in plastic boxes. To ground an outlet in a plastic box (near left), attach a jumper wire to the grounding terminal on the outlet, then twist the end of the jumper together with the bare wires entering the box and screw on a wire cap. If there is only one bare wire entering the box, connect it directly to the grounding terminal on the outlet.

SERVICING DUPLEX OUTLETS

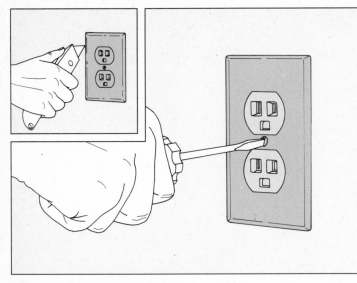

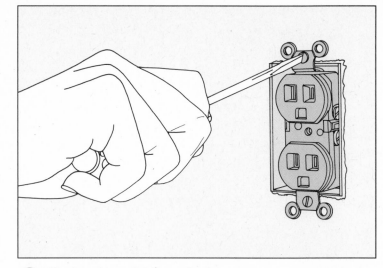

1 **Removing the cover plate.** Turn off power to the outlet by removing the fuse or tripping the circuit breaker *(page 18)*. Remove the screw on the cover plate, as shown, and lift it away from the wall. If the cover plate is stuck to the wall with paint or plaster, use a utility knife or single-edge razor blade to cut neatly around its edge *(inset)*. Tape the screw to the cover plate to avoid losing it.

2 **Freeing the outlet from the box.** Release the outlet from the box by loosening the mounting screws located at the top and bottom of the mounting strap *(above)*. Grasp the mounting strap and pull the outlet out of the box to expose the wires. Do not touch any wires or terminals until you have tested for voltage *(step 4)*.

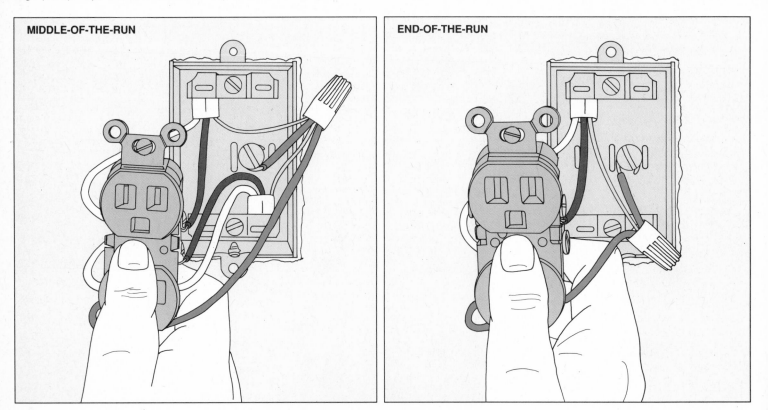

MIDDLE-OF-THE-RUN

END-OF-THE-RUN

3 **Identifying the wiring.** Check the number of cables entering the box. When the outlet is middle-of-the-run, two cables enter the box *(above, left)*, each with one black wire, one white wire, and a bare copper grounding wire. The black wires are hot (they carry current) and are connected to the brass terminals on the outlet. The white wires are neutral and are connected to the silver terminals. When the outlet is end-of-the-run, only one cable enters the box *(above, right)*. The cable contains one black wire, one white wire and a bare grounding wire. The black wire is attached to a brass terminal and the white wire is attached to a silver terminal. In both cases, a grounding jumper is attached to the grounding terminal on the outlet.

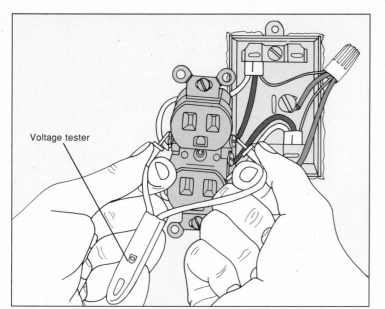

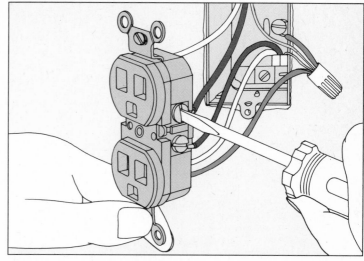

5 **Servicing the connections.** Check the connections at each terminal. If a wire is loose, rehook it around the terminal and tighten the screw. If the connections appear dirty, detach the wires and clean the terminals and wire ends with fine sandpaper. Or clip the wires and strip the insulation to expose clean wire *(page 141)*, then form each wire end into a hook and secure it to the terminal screw, as shown. Screw the outlet back into the box, turn on the power, and plug a lamp into each receptacle in turn. If the lamp doesn't work in either receptacle, turn off the power and replace the outlet *(next step)*.

4 **Testing for voltage.** Use a voltage tester to confirm that the power is off by touching one of the tester probes to a brass terminal where a wire is connected and the other probe to a silver terminal where a wire is connected *(above)*. If there are wires connected to the other set of terminals, repeat the procedure for these too. Then test between the grounding terminal and the brass and silver terminals in succession. The tester should not glow in any test. If it does, return to the service panel and turn off power to the correct circuit.

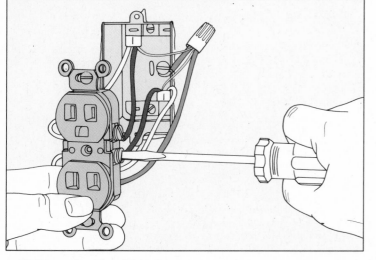

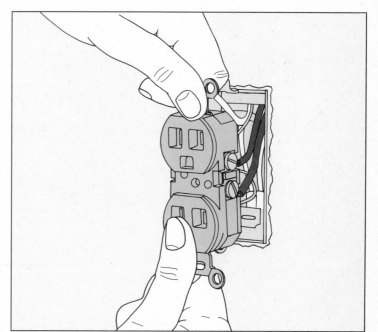

6 **Removing the old outlet.** Loosen the terminal screws and disconnect the black and white wires, as shown. Detach the grounding jumper from the green grounding terminal on the outlet.

7 **Installing the new outlet.** Loosen the terminal screws of the new outlet. Hook each black wire around a brass terminal and tighten the connections, then hook each white wire around a silver terminal and tighten the connections. Connect the grounding jumper to the green grounding terminal on the outlet. Gently fold the wires into the box and set the outlet in place *(above)*. Screw the mounting strap to the box, making sure that the outlet is straight. Put on the cover plate and turn on the power.

SERVICING SWITCH-CONTROLLED DUPLEX OUTLETS (Switch controls both receptacles)

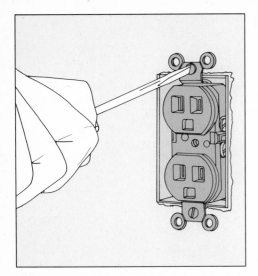

1 **Removing the outlet.** When an outlet controlled by a wall switch is not operating, first check the switch *(page 70)*. If the problem persists, service the outlet. Begin by flipping off the switch that controls the outlet, then turn off the power by removing the fuse or tripping the circuit breaker *(page 18)*. Unscrew the cover plate, loosen the mounting strap *(above)* and pull the outlet from the box, taking care not to touch any exposed wire ends or terminals.

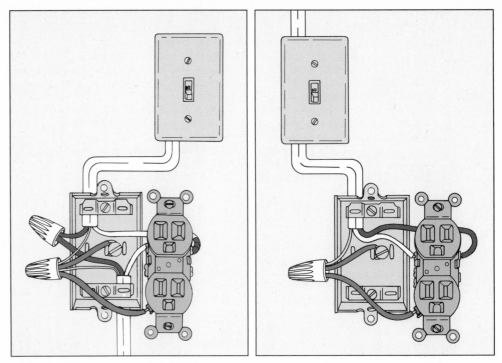

2 **Identifying the wiring.** If two cables enter the box *(above, left)*, one of the white wires is attached to a brass terminal; this white wire should be recoded black with electrical tape or black paint to indicate that it is hot (it carries current to the switch). The other white wire is attached to a silver terminal. The black wires are connected under a wire cap. If one cable enters the box *(above, right)*, the black wire is attached to a brass terminal and the white wire is attached to a silver terminal.

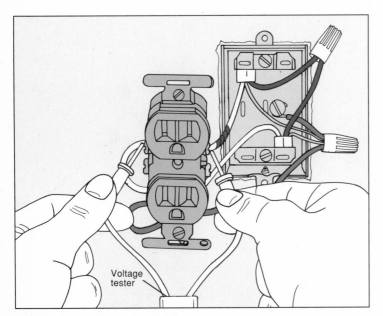

Voltage tester

3 **Testing for voltage.** Use a voltage tester to confirm that the power is off by touching one of the tester probes to the brass terminal where a wire is connected and the other probe to the silver terminal where a wire is connected *(above)*. Then test between the grounding terminal and the brass and silver terminals in succession. The tester should not glow in any test. If it does, return to the service panel and turn off power to the correct circuit.

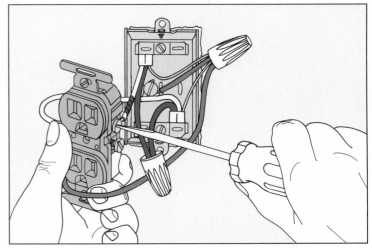

4 **Servicing the connections.** Check the connections at each terminal. If a wire is loose, rehook it around the terminal and tighten the screw. If the connections appear dirty, detach the wires and clean the terminals and wire ends with fine sandpaper. Or clip the wires and strip the insulation to expose clean wire *(page 141)*, then form each wire end into a hook and secure it to the terminal, as shown. Screw the outlet back into the box, turn on the power, flip on the wall switch and plug a lamp into each receptacle. If the lamp doesn't work in either receptacle, turn off the power, disconnect the outlet and install a replacement *(next step)*.

SERVICING SWITCH-CONTROLLED DUPLEX OUTLETS (Switch controls one receptacle)

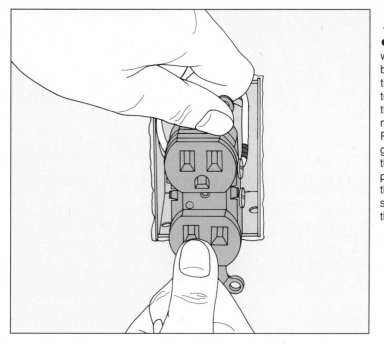

5 **Replacing the outlet.** Loosen the terminal screws on the new outlet. If there are two cables in the box, hook the white wire with the black marking around a brass terminal and tighten the connection, then hook the other white wire around a silver terminal and tighten it. If there is one cable in the box, attach the black wire to a brass terminal and the white wire to a silver terminal. Reconnect the grounding jumper to the green grounding terminal on the outlet. Gently fold the wires into the box and set the outlet in place *(left)*. Screw the mounting strap onto the box, making sure that the outlet is straight. Put on the cover plate and turn on the power.

SERVICING SWITCH-CONTROLLED OUTLETS (Switch controls one receptacle)

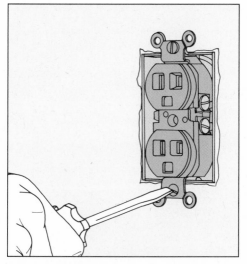

1 **Removing the outlet.** This duplex outlet is wired so that one receptacle (upper or lower) is controlled by a wall switch; the other receptacle is independent of the switch. When the outlet does not operate correctly, first check the switch *(page 70)*. If the problem persists, service the outlet. Turn off the power by removing the fuse or tripping the circuit breaker *(page 18)*. Unscrew the cover plate, loosen the mounting strap and pull the outlet from the box, taking care not to touch any exposed wire ends or terminals.

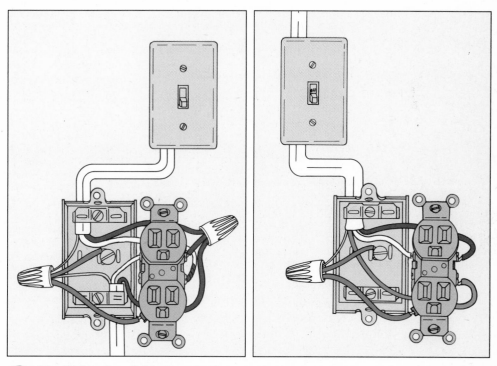

2 **Identifying the wiring.** If two cables enter the box *(above, left)*, the black wires are connected to a jumper that hooks around a brass terminal. One of the white wires is attached to a silver terminal. The other white wire is attached to a brass terminal; this white wire should be recoded black with electrical tape or black paint, indicating that it is hot (it carries current to the switch). If one cable enters the box *(above, right)*, it has two hot wires (one black and one red) to deliver power to the receptacles individually. The black and red wires are attached to brass terminals. The white wire is attached to a silver terminal. In both cases, the connecting tab between the brass terminals has been removed, interrupting the flow of power between the upper and lower receptacles.

WALL OUTLETS

SERVICING SWITCH-CONTROLLED DUPLEX OUTLETS (Switch controls one receptacle, continued)

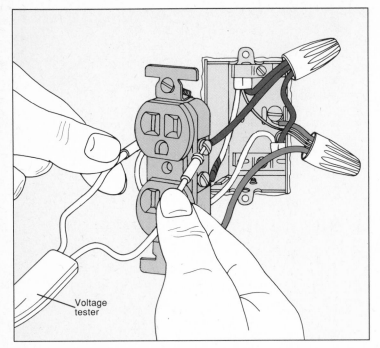

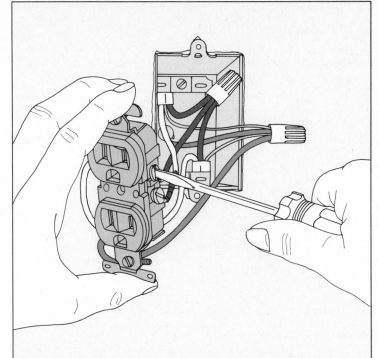

3 **Testing for voltage.** Use a voltage tester to confirm that the power is off by touching one of the tester probes to one brass terminal and the other probe to the silver terminal where the wire is connected. Repeat the procedure for the same silver terminal and the other brass terminal. Then test between the grounding terminal and the brass and silver terminals in succession. The tester should not glow in any test. If it does, return to the service panel and turn off power to the correct circuit.

Voltage tester

4 **Servicing the connections.** Check the connections at the terminals. If a wire is loose, rehook it around the terminal and tighten the screw. If the connections appear dirty, detach the wires and clean the terminals and wire ends with fine sandpaper. Or clip the wires and strip the insulation to expose clean wire *(page 141)*, then form each wire end into a hook and secure the wires back under the terminals, as shown. Screw the outlet back into the box, turn on the power, flip on the wall switch and plug a lamp into each receptacle. If the lamp doesn't work in either receptacle, turn off the power, disconnect the outlet and install a replacement *(next step).*

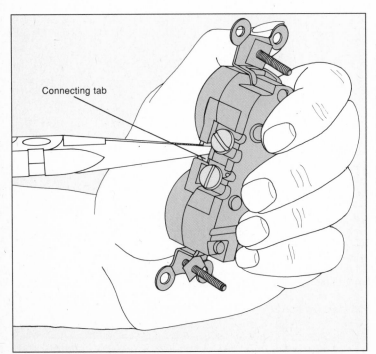

Connecting tab

5 **Installing the new outlet.** To ensure that the two receptacles of the new outlet will operate independently, use long-nose pliers or a screwdriver to snap off the connecting tab between the brass terminals *(left).* If there are two cables in the box, hook the white wire with the black marking around one brass terminal and the jumper from the pigtail of black wires around the other brass terminal and tighten the connections. Connect the white wire to a silver terminal. If there is one cable in the box, attach the black wire to one brass terminal, the red wire to the other brass terminal, and the white wire to a silver terminal. Reconnect the grounding jumper to the green grounding terminal on the outlet. Screw the outlet into the box, making sure that it is straight, then put on the cover plate and turn on the power.

SERVICING SPLIT-CIRCUIT DUPLEX OUTLETS

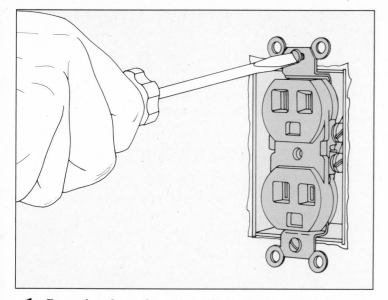

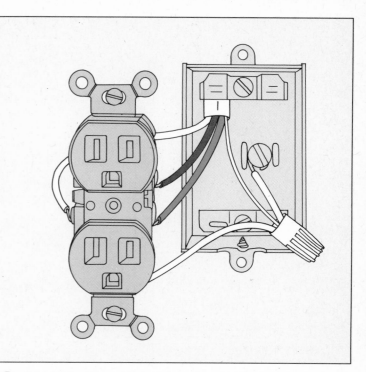

1 **Removing the outlet.** On a split-circuit duplex outlet, the upper and lower receptacles of the outlet operate independently, each connected to half of a dedicated 120/240-volt circuit. Commonly found in kitchens, the split-circuit duplex provides power for appliances, such as toasters, electric kettles and coffeemakers, which consume a great deal of electricity. When the outlet is not operating correctly, turn off power to the 120/240-volt circuit at the main service panel *(page 18)*, or at a switch beside the panel. Unscrew the cover plate, loosen the mounting strap *(above)* and pull the outlet from the box. Take care not to touch any exposed wire ends or terminals.

2 **Identifying the wiring.** One cable enters the box; it has two hot wires (one black and one red), each attached to a brass terminal. The white wire is attached to a silver terminal. The connecting tab between the brass terminals has been removed, interrupting the flow of power between the upper and lower receptacles.

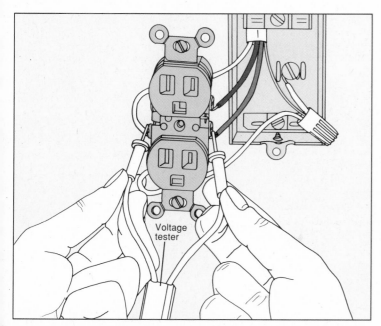

Voltage tester

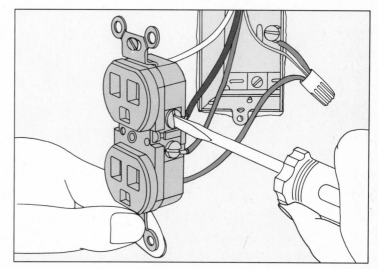

3 **Testing for voltage.** Use a voltage tester to confirm that the power is off by touching one of the tester probes to a brass terminal and the other probe to the silver terminal where a wire is connected. Repeat the procedure for the same silver terminal and the other brass terminal. Then test between the grounding terminal and the brass and silver terminals in succession. The tester should not glow in any test. If it does, return to the service panel and turn off power to the correct circuit.

4 **Servicing the connections.** Check the connections at the terminals. If a wire is loose, rehook it around the terminal and tighten the screw. If the connections appear dirty, detach the wires and clean the terminals and wire ends with fine sandpaper. Or clip the wires and strip the insulation to expose clean wire *(page 141)*, then form the wire ends into hooks and secure each wire under its terminal, as shown. Screw the outlet back into the box, turn on the power, and test the outlet. If the outlet still doesn't work, turn off the power, disconnect the outlet and install a replacement *(next step)*.

SERVICING SPLIT-CIRCUIT DUPLEX OUTLETS (continued)

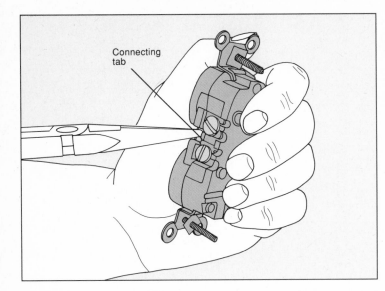

Connecting
tab

5 **Installing the new outlet.** To ensure that the upper and lower receptacles of the new outlet will operate independently, use long-nose pliers or a screwdriver to snap off the connecting tab between the brass terminals *(left)*. Attach the black wire to one brass terminal, the red wire to the other brass terminal, and the white wire to a silver terminal. Reconnect the grounding jumper to the green grounding terminal on the outlet. Gently fold the wires into the box. Screw the mounting strap onto the box, making sure that the outlet is straight, then put on the cover plate and turn on the power.

SERVICING GROUND-FAULT CIRCUIT INTERRUPTERS

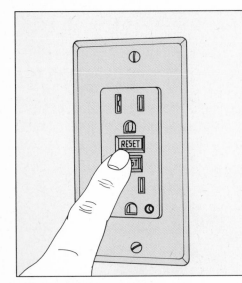

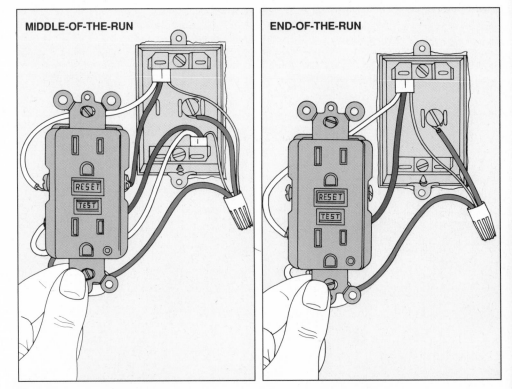

MIDDLE-OF-THE-RUN

END-OF-THE-RUN

1 **Testing the outlet.** The GFCI (ground-fault circuit interrupter) outlet protects the circuit—and you—by tripping instantly when it detects a leak in current *(page 25)*. Code now requires that GFCI outlets be installed in new and remodeled bathrooms, kitchens and garages; they are also recommended in workshops, laundry rooms and other damp locations. To ensure continued protection, check the GFCI outlet every month by pressing the TEST button; if the RESET button does not pop out, service the outlet. When a GFCI outlet is not delivering power to an appliance, the fault could be with the appliance or the outlet. Unplug the appliance and press the RESET button. If the GFCI still doesn't work, go to step 2.

2 **Identifying the wiring.** Turn off power to the outlet by removing the fuse or tripping the circuit breaker *(page 18)*. Unscrew the cover plate, loosen the mounting strap and pull the outlet from the box, taking care not to touch any exposed wire ends or terminals. If two cables enter the box *(above, left)*, each of the black wires is connected to a brass terminal and each of the white wires is connected to a silver terminal. If one cable enters the box *(above, right)*, the black wire is attached to the brass terminal marked LINE and the white wire is attached to the opposite silver terminal. Some GFCI outlets have leads in the place of terminals: two white, two black and one green grounding lead.

SERVICING GROUND-FAULT CIRCUIT INTERRUPTERS (continued)

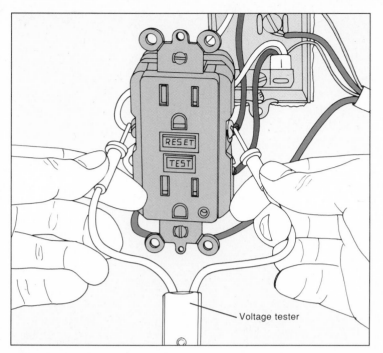

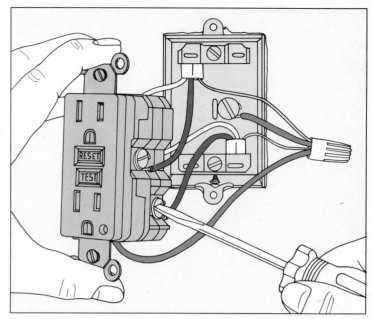

Voltage tester

3 **Testing for voltage.** Use a voltage tester to confirm that the power is off by touching one of the tester probes to the brass terminal where a wire is connected and the other probe to the silver terminal where a wire is connected *(above)*. If there are wires connected to the other set of terminals, repeat the procedure for these. Then test between the grounding terminal and the brass and silver terminals in succession. The tester should not glow in any test. If it does, return to the service panel and turn off power to the correct circuit.

4 **Servicing the connections.** Check the connections at the terminals. If a wire is loose, rehook it around the terminal and tighten the screw. If the connections appear dirty, detach the wires and clean the terminals and wire ends with fine sandpaper. Or clip the wires and strip the insulation to expose clean wire *(page 141)*, then form the wire ends into hooks and secure each wire back under its terminal, as shown. Screw the outlet back into the box, turn on the power and press the RESET button. If the outlet still doesn't work, turn off the power, free the outlet from the box and go to step 5.

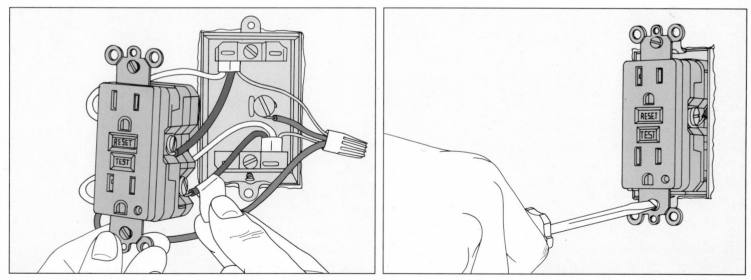

5 **Replacing the outlet.** If there are two cables in the box, tag the wires to help you when reconnecting the new outlet. Wrap masking tape around the black and white wires connected to the terminals marked LINE *(above, left)*. Disconnect the outlet and buy a replacement. If there are two cables in the box, hook the tagged black wire around the brass terminal marked LINE, and connect the tagged white wire to the opposite silver terminal. Then connect the second black wire to the second brass terminal and the second white wire to the second silver terminal. If there is one cable in the box,

attach the black wire to the brass terminal marked LINE and the white wire to the opposite silver terminal. Reconnect the grounding jumper to the green grounding terminal on the GFCI outlet. Gently fold the wires back into the box and screw the mounting strap to the box *(above, right)*, making sure that the outlet is straight. Put on the cover plate, turn on the power and press the RESET button. If the outlet still doesn't work and there are two cables in the box, a faulty appliance or connection at another point along the circuit may be causing the GFCI to trip. Check the other devices on the same circuit *(page 104)*.

REPLACING A TWO-SLOT DUPLEX OUTLET

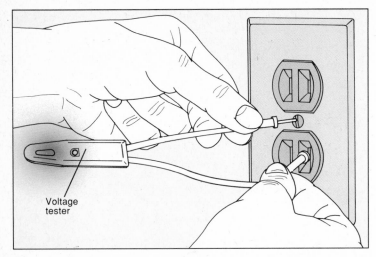

Voltage tester

1 **Testing for grounding.** To determine whether you can replace a two-slot outlet with a three-slot grounded outlet, first confirm that the outlet box is grounded. (**Caution:** This is a live voltage test. Be sure to hold the voltage tester by its insulated handles, or use the one-hand technique shown on page 134.) With the power on, place one probe of the voltage tester on the cover plate screw and insert the other probe into one slot in the outlet *(above)*, then into the other slot. The tester should glow when the probe is in one slot (the hot slot) and not in the other. If it doesn't glow at all, the cover plate is not grounded; replace it with a GFCI outlet or a two-slot outlet, or have an electrician extend a grounded circuit to the box.

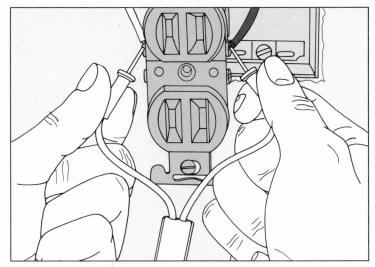

2 **Testing for voltage.** Turn off power to the outlet by removing the fuse or tripping the circuit breaker *(page 18)*. Free the outlet from the box, taking care not to touch any exposed wire ends or terminals. Use a voltage tester to confirm that the power is off by touching one probe to a brass terminal where a wire is connected and the other probe to the silver terminal on the opposite side. Repeat the test for the other brass and silver terminals if wires are connected to them. Then test between the grounded metal box and the brass and silver terminals in succession. The tester should not glow in any test. If it does, return to the service panel and turn off power to the correct circuit.

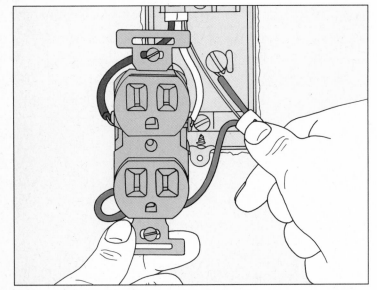

3 **Replacing the outlet.** Disconnect the old outlet and buy a three-slot replacement outlet. To attach the new outlet, hook the black wire around a brass terminal and the white wire around a silver terminal, then tighten the connections. Next, disconnect the grounding wire from the grounding screw at the back of the box. Cut two jumper wires *(page 141)*. Hook one jumper around the grounding screw at the back of the box, and the other around the grounding terminal on the new outlet, then twist together the bare grounding wire with the jumpers and screw on a wire cap, as shown. (Some armored cable has no grounding wire; to ground the outlet, attach a jumper to the screw at the back of the box and hook the other end around the grounding terminal on the outlet.) Screw the outlet into the box, put on the cover plate and turn on the power.

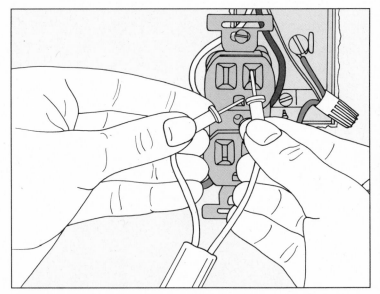

4 **Checking the grounding.** Use the voltage tester to confirm that the outlet is properly grounded by inserting one tester probe into the U-shaped grounding slot and the other into the long (neutral) slot, then into the shorter (hot) slot, as shown. (**Caution:** This is a live voltage test. Be sure to hold the tester by its insulated handles, or use the one-hand technique shown on page 134.) The tester should glow when the probe is in the hot slot and not when it is in the neutral slot. If the outlet fails the test, turn off the power and check the grounding connections.

INSTALLING A GROUNDING-ADAPTER PLUG

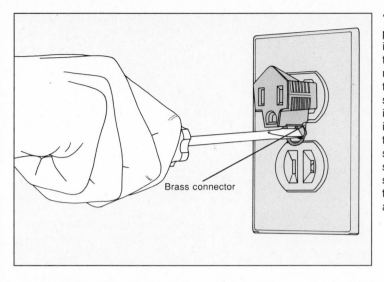

Brass connector

Temporary grounding for a two-slot outlet. A grounding-adapter plug enables a two-slot outlet to accept three-prong plugs. Before installing one, test to confirm that the outlet box is grounded, following the procedure on page 98, step 1. If the box is grounded, loosen the mounting screw on the cover plate and slip the brass connector under the screw. Use a voltage tester to confirm that the mounting screw is still grounded by touching one probe of a voltage tester to the mounting screw and inserting the other probe first into one outlet slot, then into the other slot. (**Caution:** This is a live voltage test. Be sure to hold the tester by its insulated handles, or use the one-hand technique shown on page 134.) If the tester glows when the probe contacts one slot but not the other, push the grounding-adapter plug into the outlet slots and tighten the screw to secure the connection, as shown. If the tester doesn't glow, replace the mounting screw with a longer screw and repeat the test before plugging in the adapter plug.

240-VOLT OUTLETS

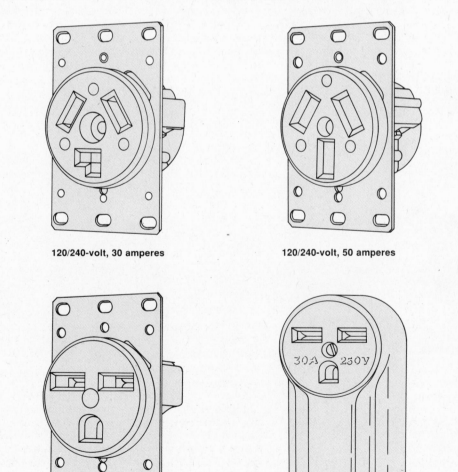

120/240-volt, 30 amperes

120/240-volt, 50 amperes

240-volt, 30 amperes

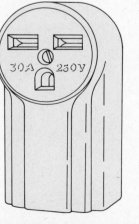

Surface-mounted

High-voltage outlets. There are two basic types of outlets with 240-volt capacity: the combination 120/240-volt outlet and the exclusively 240-volt outlet. These outlets and their plugs come as matched sets; an outlet will not admit the plug of an appliance with a different voltage and amperage rating. When replacing a 240-volt outlet, always install one of the same voltage and amperage ratings.

The 120/240-volt, 30-ampere outlet is designed to service clothes dryers, whose heating element operates on 240 volts and timer and motor operate on 120 volts. Electric ranges use a combination 120/240 outlet that carries a current of 50 amperes. The burners and oven, when set at high temperatures, require the full 240 volts; range clocks and lights run on only 120 volts. The dedicated 240-volt outlet is used for appliances with heavy-duty power requirements, such as water heaters and air conditioners. Both 120/240-volt and 240-volt outlets are available in surface-mounted models with the outlet and box in one unit. These are handy in basements or utility rooms, where flush mounting is not necessary.

Safety precautions are important when working on a 240-volt outlet. Make sure the floor is dry and wear rubber-soled shoes. Be sure to turn off the power at the service panel and test to confirm it is off before beginning any repairs. When performing a live voltage test, proceed with caution.

SERVICING 120/240-VOLT OUTLETS

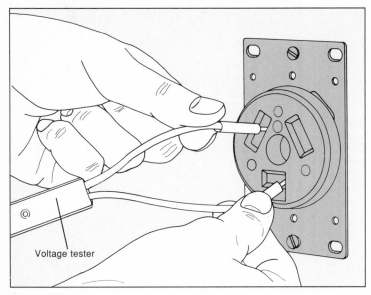

Voltage tester

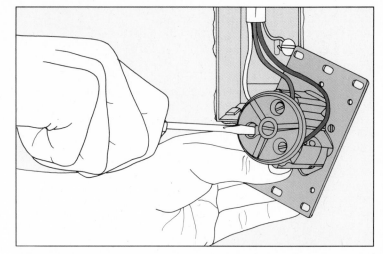

1 Testing for voltage. Turn off power to the outlet by removing the fuse block or tripping the double circuit breaker *(page 18)*. Use a voltage tester rated at over 240 volts to confirm that the power is off. Insert one probe into the lower (neutral) slot, in this case the L-shaped slot, and the other probe into each of the upper slots in turn *(above)*. Then test between the two upper slots. Hold the probes by their insulated handles only and push them into the outlet, so that they contact the metal parts inside the slots. (If the probes do not contact the metal parts, use a multitester or a bar-meter voltage tester with longer probes.) The tester should not glow in any test. If it does, return to the service panel and turn off power to the correct circuit.

2 Servicing the connections. With the power confirmed off, loosen the screw holding the cover plate in place, if there is one, and remove it. Release the outlet from the box by loosening the screws on the top and bottom of the mounting plate. Grasp the mounting plate and pull the outlet out of the box. Check the connections at the terminals. If they appear dirty or corroded, detach the wires and clean the wire ends with fine sandpaper. Or clip the wires and strip the insulation to expose clean wire *(page 141)*. Then push each wire back into its terminal hole, making sure no uninsulated wire is exposed. If necessary, clip the wire ends to ensure a proper fit. Tighten the retaining screws to secure the connections *(above)*. Screw the outlet back into the box and turn on the power.

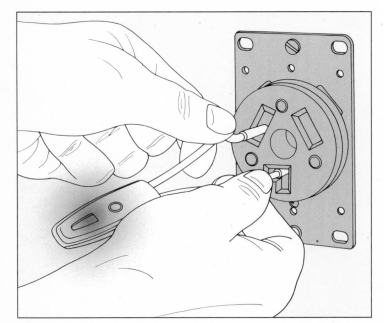

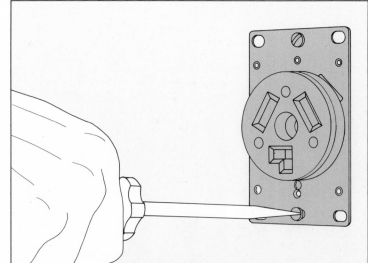

3 Testing the outlet. Use the voltage tester to confirm that the outlet is delivering power. (**Caution:** This is a live voltage test. Be sure to hold the tester by its insulated handles, or use the one-hand technique shown on page 134.) Insert one tester probe into the lower (neutral) slot, in this case the L-shaped slot, and the second probe into each of the upper slots in succession. The tester should glow in both cases. If it doesn't, turn off the power again and replace the outlet *(next step)*.

4 Replacing the outlet. Disconnect the outlet and buy a replacement that matches the voltage and amperage of the original. Hold the new outlet so that the neutral slot is at the bottom. Push the white wire into the silver terminal (or the terminal marked WHITE), then tighten the retaining screw. Push the red wire into one brass terminal hole and the black wire into the other brass terminal hole and tighten the retaining screws. Set the outlet into the box and screw on the mounting plate, as shown. Turn on the power and plug in the appliance.

SERVICING 240-VOLT OUTLETS

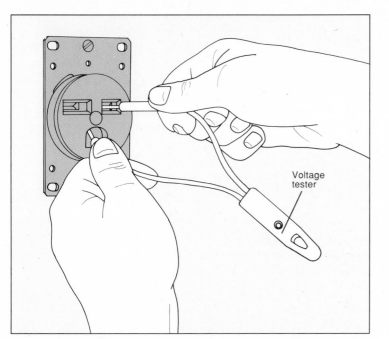

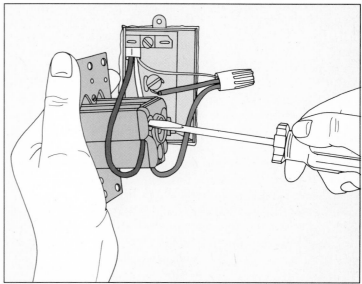

1 Testing for voltage. Turn off power to the outlet by removing the fuse block or tripping the double circuit breaker *(page 18)*. Use a voltage tester rated at 240 volts to confirm that the power is off by inserting one probe into the U-shaped grounding slot and the second probe into each of the upper slots in turn *(above)*. Then test between the two upper slots. Hold the probes by their insulated handles only and push them into the outlet so that they contact the metal parts inside the slots. (If the probes do not contact the metal parts, use a multitester or a bar-meter voltage tester with longer probes.) The tester should not glow in any test. If it does, return to the service panel and turn off power to the correct circuit.

2 Servicing the connections. Once the power is confirmed off, loosen the screw holding the cover plate in place, if there is one, and remove it. Release the outlet from the box by loosening the screws on the top and bottom of the mounting plate. Grasp the mounting plate and pull the outlet out of the box. Check the connections at the terminals. If they appear dirty or corroded, detach the wires and clean the wire ends with fine sandpaper. Or clip the wires and strip the insulation to expose clean wire *(page 141)*. Then push each wire back into its terminal hole, making sure no uninsulated wire is exposed. If necessary, clip the wire ends to ensure a proper fit. Tighten the retaining screws to secure the connections *(above)*, then screw the outlet back into the box and turn on the power.

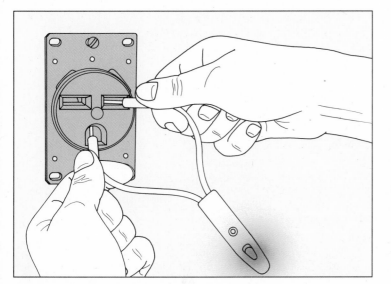

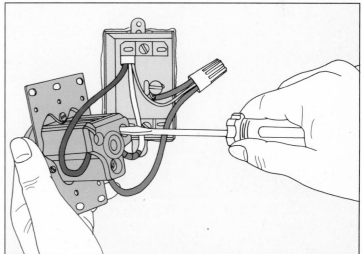

3 Testing the outlet. Use the voltage tester to confirm that the outlet is delivering power. (**Caution:** This is a live voltage test. Be sure to hold the tester by its insulated handles, or use the one-hand technique shown on page 134.) Insert one tester probe into the U-shaped grounding slot and the second probe into one of the upper slots *(above)*, then into the other. The tester should glow in both tests. If it doesn't, turn off the power and replace the outlet *(next step)*.

4 Replacing the outlet. Disconnect the outlet and buy a replacement that matches the voltage and amperage of the original. Hold the new outlet so that the grounding slot is at the bottom. Push the black wire into one brass terminal hole and the white wire into the other brass terminal hole. Push the jumper from the grounding pigtail into the green grounding terminal hole (where GR is stamped). Tighten the retaining screws *(above)*. Screw the outlet into the box. Turn on the power and plug in the appliance.

WIRING WITHIN WALLS

When an entire circuit fails, nine times out of ten the cause will be found in an electrical box. This problem can be tracked down by inspecting each outlet, switch and fixture along the circuit. Once you have pinpointed the loose connection or broken wire, the repair is usually simple *(page l04)*.

But occasionally the problem lies in the wiring behind the wall. The most common cause of damaged cables here is nailing or drilling into a wall. If the power fails—even on a single circuit—while you are hanging a picture or installing shelves, you can suspect that a cable has been pierced. Sometimes there are no clues to lead you to the damage. In either case, the troubleshooting is best left to an electrician.

However, you can extend a circuit that is inadequate for your needs. If there are not enough outlets in a room, the outlets are in the wrong location, or you want to add a new fixture to provide more light, you can tap into an existing box and run cable from it to a new location.

In planning an extension, first make sure that the existing circuit can carry the additional load *(page 23)*. Choose a box that is close to the new location, or that will provide the easiest route for the cable. Outlet boxes usually provide a better source for a circuit extension than switch boxes. The wire connections at an existing outlet are usually simpler, and running a cable from an outlet near the floor often causes less visible damage to the wall than extending it from a switch box halfway up the wall. Remember that if you extend a circuit from a switch box with only one cable, the extension will be controlled by that switch. If you extend a circuit from a switch box with two cables, you must first identify the black wire that carries the current into the box *(page 119)*.

Although the technique is not difficult, running new cable can be a time-consuming process. When you plan your circuit extension, try to make use of accessible areas. It is faster and less messy to run the cable through an unfinished basement or

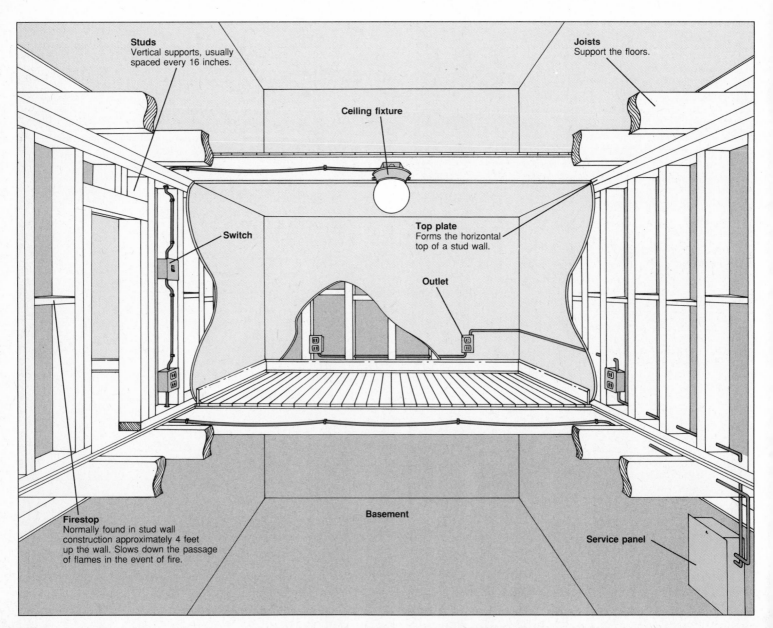

Studs
Vertical supports, usually spaced every 16 inches.

Joists
Support the floors.

Ceiling fixture

Switch

Top plate
Forms the horizontal top of a stud wall.

Outlet

Firestop
Normally found in stud wall construction approximately 4 feet up the wall. Slows down the passage of flames in the event of fire.

Basement

Service panel

attic than through finished walls that must later be patched and painted. If possible, choose a box that will allow the cable to travel along the side of basement or attic joists rather than through them. If the circuit extension to a new outlet is short, you may be able to remove the baseboard and conceal the wiring behind it.

Cable is "fished" through walls and ceilings with long, flexible metal tapes *(page 132)*. The tapes are guided through the empty space within walls, hooked to electrical cable, and pulled from the existing box to the new location. In most cases, you will need two fish tapes: one to travel through the walls and the other to pull out or "fish" the first one. For most of the work in fishing cable, you will also need a helper.

When measuring the amount of cable you will need, include the drop from the existing box to the path it will travel, add 20 percent to allow for slack in the cable, then add two or three feet for good measure.

Use a multipurpose tool or cable stripper to determine the gauge of the wire in the existing box and be sure that the new cable matches that gauge. If you run cable through a notch or along a channel in a stud or joist, protect it from being pierced by covering it with a metal plate *(page 132)*. Secure the cable to a wooden support every 4 1/2 feet along the run.

If you are extending a circuit in a workroom or laundry room, you can save time by using raceway—surface wiring that is easy to install but less attractive than hidden wiring. One type consists of metal or plastic channels mounted directly on the wall. Cable or wire is fished through these channels without having to drill through walls or supports.

Before opening an electrical box, turn off power to the circuit at the main service panel and use a voltage tester to confirm that the power is off. Familiarize yourself with the voltage tests and wiring for outlets *(page 86)*, switches *(page 70)* and fixtures *(page 43)*.

TROUBLESHOOTING GUIDE

SYMPTOM	POSSIBLE CAUSE	PROCEDURE
Fuse blows or circuit breaker trips	Circuit overloaded	Lessen the load by moving appliances to a different circuit Call an electrician to add a new circuit
	Short circuit in appliance	Inspect cords and plugs of appliances on circuit; repair if necessary
	Short circuit in box	Inspect switch, outlet and fixture boxes on circuit; repair wiring *(p. 104)* □◗
	Short circuit in cable behind wall	Call an electrician to locate and repair damage
Switch, outlet or fixture fails to work or works intermittently	Loose connection in box	Tighten connections *(switches, p. 70; outlets, p. 86; fixtures, p. 43)*
	Switch, outlet or fixture faulty	Test and replace *(switches, p. 70; outlets, p. 86; fixtures, p. 43)*
Switch or outlet hot to the touch or sparking	Dirty or loose connection in box	Clean and tighten connections *(switches, p. 70; outlets, p. 86)*
More than one switch, outlet or fixture not operating	Wire disconnected in a box on circuit	Locate and reconnect wire *(p. 104)* □◗
	Broken cable behind wall	Call an electrician to locate and repair damage
Several circuits not working	Problem at service panel	Call an electrician
No circuits working	Power failure	Turn off major appliances *(p. 9)*
	Problem at service panel	Call an electrician
Not enough outlets, switches or fixtures in room	Insufficient wiring	Extend existing circuit: Through the basement *(p. 105)* ■●▲ Through the attic *(p. 107)* ■●▲ Behind a baseboard *(p. 109)* ■●▲ Behind the wall *(p. 110)* ■●▲ Through the ceiling *(p. 110)* ■●▲ Install raceway *(p. 112)* ◖◗▲ Call an electrician to add a new circuit

DEGREE OF DIFFICULTY: □ Easy ◖ Moderate ■ Complex
ESTIMATED TIME: ○ Less than 1 hour ◗ 1 to 3 hours ● Over 3 hours   ▲ Fish tapes required

REPAIRS WITHIN BOXES

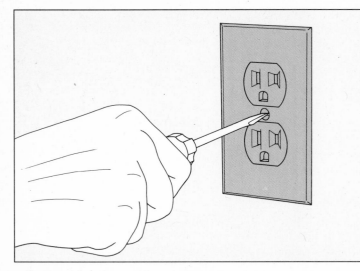

1 **Locating the problem.** When a fuse blows or circuit breaker trips constantly, the problem may be an overloaded circuit. Moving an appliance to another circuit is the easiest solution; another is to have an electrician add a circuit to your system. The problem may also be traced to a faulty appliance or lamp; inspect all the plugs and cords connected to that circuit. Most failures are caused by faulty wiring in an electrical box. Before inspecting the boxes along the circuit, familiarize yourself with the wiring in outlet boxes *(page 86)*, switch boxes *(page 70)* and fixture boxes *(page 43)*. Turn off power to the circuit by removing the fuse or tripping the circuit breaker *(page 18)*. Follow the instructions in the outlets, switches and fixtures sections to confirm that the power is off. Next, inspect the wiring in the boxes, one by one, until you have located the problem. Begin with the outlets and switches on the circuit, then check hard-to-reach fixtures such as chandeliers. To reconnect a loose wire, go to step 2. To repair a damaged wire, go to step 3.

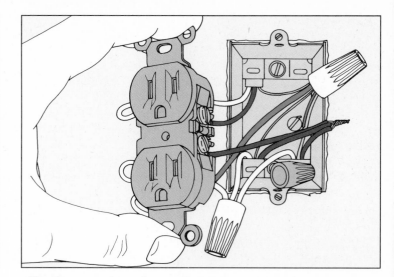

2 **Reconnecting a loose wire.** When a wire works loose from a wire cap *(above)* or terminal screw, and its bare metal end contacts the box or another bare wire, a short circuit results, blowing the fuse or tripping the circuit breaker. When a wire works loose from a connection but does not touch the box or another wire end, the current will not flow beyond that point. To reinstall a wire cap connection, make sure that the wire ends are firmly twisted together, then screw on a wire cap of the correct size. Reconnect a detached wire by hooking the end around its terminal and tightening the screw. Gently fold the wires back into the box.

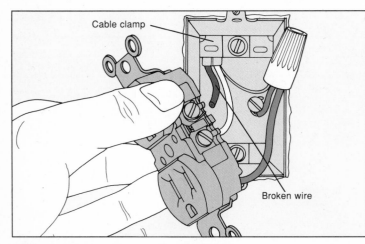

Cable clamp

Broken wire

3 **Repairing a broken wire.** A broken wire that is touching the metal box or another wire end will cause a short circuit. Loosen the terminal screw holding the broken wire and unhook the wire from the outlet or switch. Clip back the wire to remove the damage, then strip back the insulation, taking care not to nick or gouge the wire *(page 141)*. If the clipped wire is too short to reach the terminal, loosen the cable clamp in the box and pull in some slack cable, or go to step 4 to attach an extension wire. Use long-nose pliers to curl the wire end, then hook it around the terminal and tighten securely. Gently fold the wires back into the box.

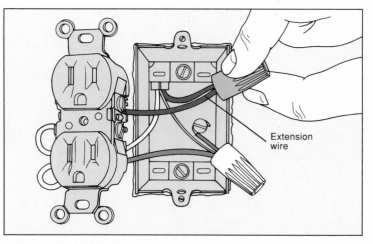

Extension wire

4 **Using an extension wire.** If the clipped wire end is too short to reconnect to the switch or outlet, make an extension wire. Measure the gauge of the shortened wire with a multipurpose tool or cable stripper and use wire of a matching gauge. Cut a piece long enough to reach from the end of the broken wire to the terminal, then strip back the insulation on both ends. Curl one end of the extension wire into a hook and attach it to the outlet or switch terminal. Twist the other end together with the shortened house wire and screw on a wire cap *(above)*.

RUNNING CABLE THROUGH THE BASEMENT

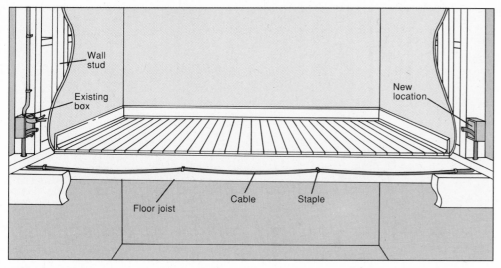

Wall stud

Existing box

New location

Floor joist

Cable

Staple

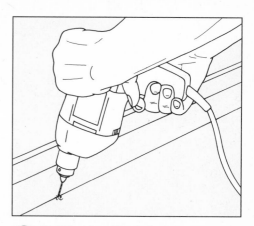

1 **Planning the run.** The ideal way to extend a circuit on a ground floor is to fish cable down into an unfinished basement, run it along or through floor joists, then feed it up to the new location. To begin, turn off the power to the circuit you are extending by removing the fuse or tripping the circuit breaker *(page 18)*. Then perform a voltage test at the existing box to confirm that the power is off. Pull the outlet or switch from the wall and disconnect it. Consult the chart on page 118 to determine whether the box is large enough to accommodate additional wires. Remove the internal clamp at the bottom of the box and punch or pry out one of the knockouts behind the clamp. Mark the location of the new box *(page 115)* and cut its hole. To calculate the length of cable you will need, measure the distance from the existing box to the hole, taking into account the distance down into the basement and back up, then add 20 percent. You will need two fish tapes and a helper to run the cable.

2 **Drilling location holes.** At the existing box, use a utility bar to pry off a section of the quarter-round molding, if any, that covers the joint between the baseboard and the floor. To help you locate the existing box while you are working in the basement, drill a 1/4-inch location hole through the floor directly under the box, as shown. You may have to use an extension bit on the drill to make the hole through to the basement. Insert a stiff piece of wire into the hole so that it is visible in the basement. Repeat the procedure directly under the location of the new box.

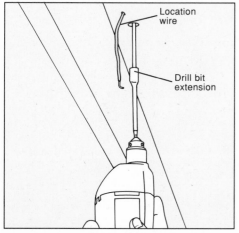

Location wire

Drill bit extension

3 **Drilling cable holes.** Find the location wires in the basement. Remember that the location holes were drilled an inch or so outside the walls. You will want to drill the 3/4-inch cable holes inside the walls. (A good guideline is 2 inches in from the location wires). Drill up through the floor, using a 3/4-inch spade bit and, when the bit is not long enough, an extension *(above)*.

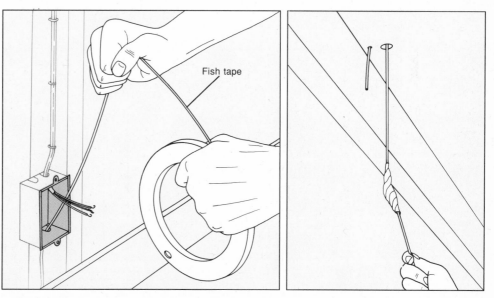

Fish tape

4 **Feeding fish tape through the existing box.** Have a helper in the basement guide a fish tape up through the hole leading to the existing box, while you listen for it at the box. Spread open the hook of the second fish tape, then feed it through the knockout hole in the bottom of the box. Move your tape back and forth in the wall cavity until you snag the fish tape from the basement, then pull both tapes up through the knockout hole *(above, left)*. Secure the two fish tapes by pinching the hooks closed and taping the connection. Then have your helper pull both fish tapes down through the floor into the basement *(above, right)*.

RUNNING CABLE THROUGH THE BASEMENT (continued)

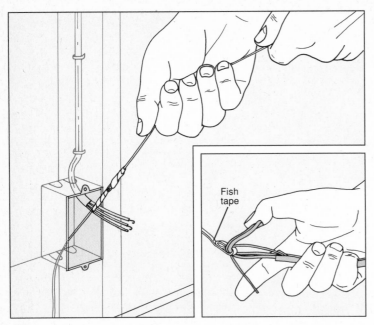

5 **Fishing the new cable to the existing box.** In the basement, detach the fish tapes and set aside the tape that was used from below. Strip back 8 inches of sheathing from the new cable *(page 138)*, thread the wires through the hook *(inset)*, then tape the connection. Have a helper feed the cable up from the basement as you pull the fish tape up through the existing box *(above)*. Pull the stripped cable into the box, detach it from the fish tape and secure the cable to the box by tightening the internal clamp.

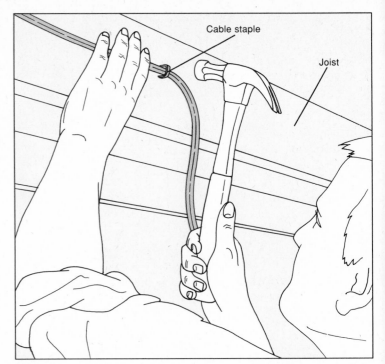

6 **Running the cable along a joist.** Whenever possible, run the cable along the side of a wooden joist, securing it every 4 1/2 feet with cable staples, as shown. Be careful not to damage the cable as you are hammering in the staples; it takes very little pressure to cut into plastic sheathing.

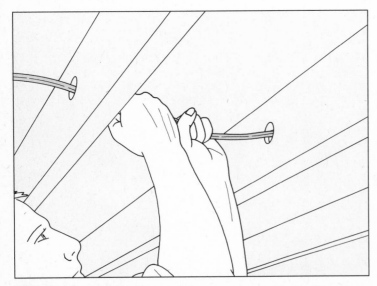

7 **Running cable through joists.** Sometimes you may have to run the cable through several joists. Use a 3/4-inch spade bit to drill a series of holes for the cable. Drill through the center of the joist; cutting too close to the top or bottom will weaken it. Pass the cable through each joist, as shown.

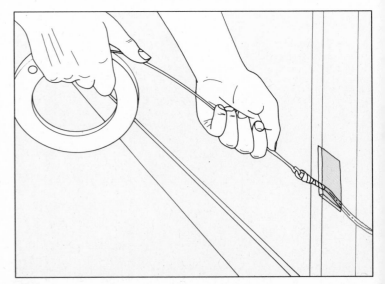

8 **Fishing cable to the new location.** Repeat steps 4 and 5 to fish the cable to the new location. The job will be easier since you are working through a hole in the wall *(above)* rather than a knockout in an existing box. Once the cable has been pulled through the hole at the new location, cut it, allowing 8 inches to make the wire connections. Install the new box *(page 115)* and connect the new outlet or switch *(page 118)*. Then go to the existing box and connect the cable to the old switch or outlet *(page 118)*.

RUNNING CABLE THROUGH THE ATTIC

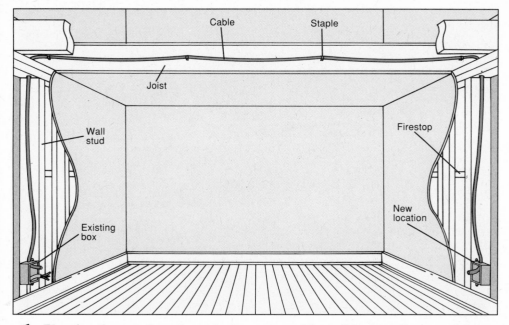

Cable   Staple

Joist

Wall stud

Firestop

New location

Existing box

1 **Planning the run.** Extending a circuit on a second floor will be faster if you can fish from an existing circuit up to an unfinished attic, run the cable along the attic floor joists, then feed it down to the new location. Turn off the power to the circuit you are extending by removing the fuse or tripping the circuit breaker *(page 18)*. Then perform a voltage test at the existing box to confirm that the power is off. Pull the outlet or switch from the wall and disconnect it. Consult the chart on page 118 to determine whether the box is large enough to accommodate additional wires. Remove the internal clamp at the top of the box and punch or pry out one of the knockouts behind the clamp. Mark the location of the new box *(page 115)* and cut its hole. Measure the amount of cable you will need, including the distance up to the attic and back down, then add 20 percent. You will need two fish tapes and a helper to run the new cable.

2 **Drilling location and cable holes.** Drill a 1/4-inch location hole through the ceiling directly above the existing box, as shown; if the drill bit is not long enough, use an extension. Insert a stiff wire into the hole so that its end is visible in the attic. Repeat the procedure directly above the new location. In the attic, lay planks across the exposed joists so that you can move around more easily. Find the location wires. Remember that the location holes were drilled an inch or so outside the walls. You will want to drill 3/4-inch cable holes inside the wall cavity. (A good guideline is 2 inches in from the location wire). Drill down through the top plate of the wall using a 3/4-inch spade bit and, when the bit is not long enough, an extension.

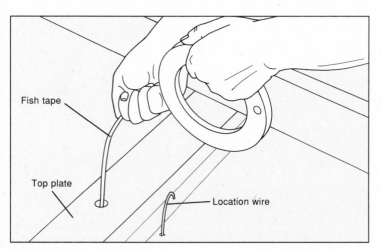

Fish tape

Top plate

Location wire

3 **Feeding the fish tape down from the attic.** Working from the attic, feed the fish tape down through the hole above the existing box, as shown. Have a helper on the floor below listen for the fish tape. If the tape catches on an obstruction, shake or snap it free, then continue feeding it down. If the fish tape becomes blocked halfway down the wall, it has probably hit a firestop; go to step 4. If it reaches the existing box without snagging, continue to step 6.

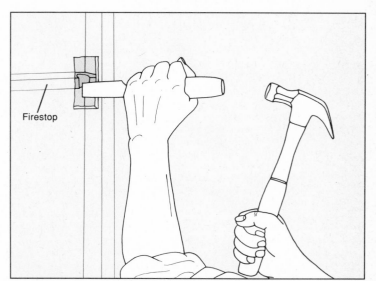

Firestop

4 **Making a passage through a firestop.** If the fish tape is blocked by a firestop—a horizontal piece between the studs—have your helper listen to locate and mark the spot where the fish tape stops. With a utility knife, score the outline of a 3- by 5-inch hole at the firestop. Use a hammer and chisel to remove the drywall, then chisel a notch in the wood *(above)* to create a passage for the fish tape and cable. Patch the hole after you have installed the cable.

RUNNING CABLE THROUGH THE ATTIC (continued)

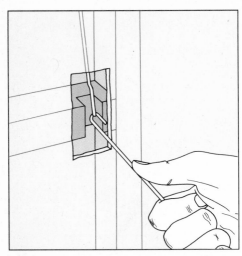

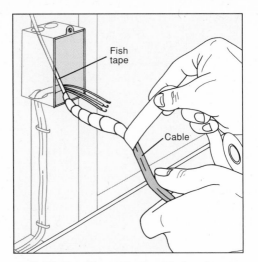

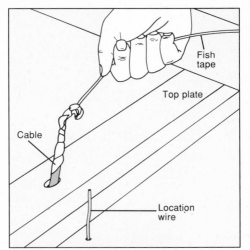

5 Fishing through the firestop. Working from the attic, move the fish tape back and forth along the firestop as your helper inserts a second fish tape with an opened hook to snag your tape. When it is caught, have the helper pull it out of the wall *(above)* and pass it through the notch past the firestop. Then continue feeding the first tape down toward the existing box. After the cable is installed, protect it by covering the firestop notch with a metal plate *(page 132)*.

6 Fishing through the existing box. Have the helper feed the second fish tape (with an opened hook) through the knockout hole in the top of the box while you twist the first fish tape back and forth from the attic. When the second fish tape has snagged the first tape, have the helper pull both tapes through the box and disconnect them. Strip back 8 inches of sheathing from the new cable *(page 138)*, thread the wires through the hook and tape the connection securely, as shown.

7 Pulling up the new cable. Return to the attic and pull up the fish tape while your helper feeds the cable through the knockout in the existing box. If the cable becomes stuck on its way up, twist the fish tape to free it, or pull up and down to maneuver it around the obstruction. When it reaches the attic *(above)*, haul up enough cable to cross the attic and go down to the new location, leaving 8 inches of cable at the existing box. Fasten the cable to the existing box using the internal clamp.

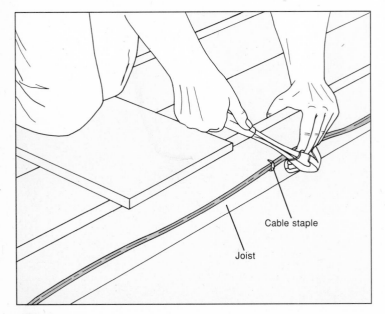

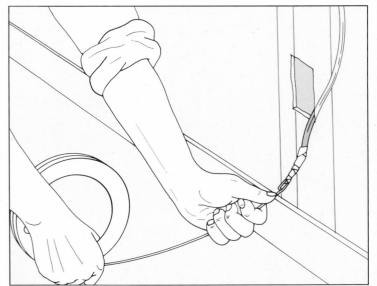

8 Running cable in the attic. Run the cable along the center of an attic joist, securing it every 4 1/2 feet with cable staples, as shown. Be careful not to damage the cable as you are hammering in the staples. If you have to run cable across the joists instead of along them, use a 3/4-inch spade bit to drill holes through the joists. Drill through the center of the joist; cutting too close to the top or bottom will weaken it. Pass the cable through each joist.

9 Running cable down to the new location. Repeat steps 3, 4 and 5 to fish the cable behind the wall to the new location. The job will be easier since you are working through a hole in the wall *(above)* rather than a knockout in an existing box. Once the cable has been pulled through the hole at the new location, install the new box *(page 115)* and connect the new outlet or switch *(page 118)*. Then go to the existing box and connect the old outlet or switch *(page 118)*. Patch any holes that you made at the firestop with scrap drywall and joint compound.

RUNNING CABLE BEHIND A BASEBOARD

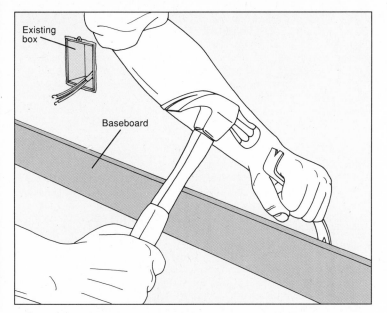

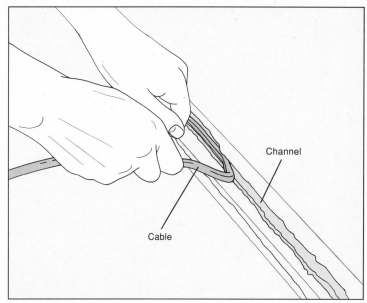

1 **Prying off the baseboard.** If your circuit extension is not too long and the baseboard can be easily removed, you can conceal the cable behind it. First turn off power to the circuit by removing the fuse or tripping the circuit breaker *(page 18)*. Then test for voltage to confirm that the power is off. Pull the outlet or switch from the wall and disconnect it. Consult the chart on page 118 to determine whether the box is large enough to accommodate additional wires. Mark the location of the new box *(page 115)* and cut its hole. To calculate the amount of cable you will need, measure from the existing box to the new location, then add 20 percent. With a utility bar, pry off the quarter-round molding between the existing box and the new location. Then pry off the baseboard, taking care not to gouge the baseboard or wall.

2 **Running the new cable.** Use a chisel or utility knife to cut a channel for the cable in the wall where you have removed the baseboard. To avoid piercing the cable when you nail the baseboard back in place, position the channel half the height of the baseboard. Then run the cable from the existing box to the new location, pushing it into the channel as you go *(above)*. Secure the cable to each stud with a cable staple.

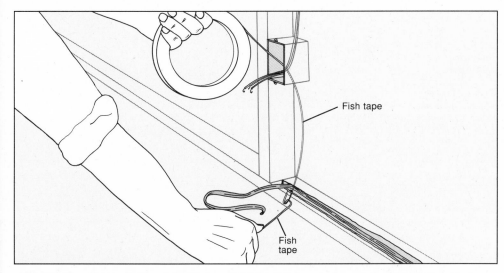

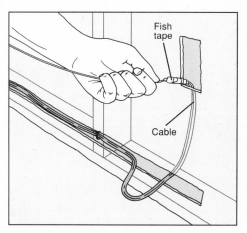

3 **Fishing through the existing box.** Feed one fish tape down into the wall through the knockout in the box until it strikes the floor. Insert the second fish tape (with an opened hook) into the channel below the box. Move the two fish tapes back and forth until they meet, then use the second fish tape to pull the first tape out of the wall at the channel, as shown. Disconnect the two fish tapes and set the second one aside. Strip back 8 inches of sheathing from the new cable, thread the wires through the hook of the fish tape, then tape the connection. Pull on the fish tape from the box, bringing the stripped cable end through the knockout. Disconnect the fish tape, then use the internal clamp to fasten the cable to the box.

4 **Fishing cable to the new location.** Lower one fish tape into the hole for the new box until it reaches the floor, feed the second fish tape into the channel and pull the cable through the hole as in step 3. Cut the cable, allowing 8 inches to make the wire connections. Install the new box *(page 115)* and connect the new switch or outlet. Then go to the existing box and reconnect the old switch or outlet *(page 118)*.

RUNNING CABLE BEHIND THE WALL

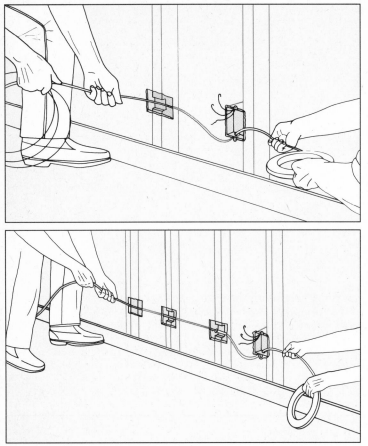

Fishing through wall studs. Since you have to break into the wall at each stud to run cable horizontally behind it, this technique should be used only as a last resort. Turn off the power to the circuit you are extending by removing the fuse or tripping the circuit breaker *(page 18)*. Then test for voltage at the existing box to confirm that the power is off. Pull the outlet or switch from the wall and disconnect it. Consult the chart on page 118 to determine whether the box is large enough to accommodate additional wires. Remove one of the internal clamps in the box and punch or pry out a knockout behind the clamp. Mark the location of the new box *(page 115)* and cut its hole. Mark a straight line from the existing box to the new location and measure the amount of cable you need. Use a stud finder *(page 132)* or drill small pilot holes to locate the vertical studs along the wall. Cut a hole through the wall on both sides of each stud, then chisel a notch across each stud large enough for the cable.

You will need two fish tapes and a helper. Have the helper guide the first fish tape through the existing box toward the nearest stud. Insert the second fish tape with the opened hook into the hole at that stud, directing it toward the first fish tape. Move it back and forth until it catches the first fish tape, then pull the first tape through the stud *(left, top)*. Separate the fish tapes, then fish through the second stud in the same way, repeating the procedure until the first fish tape has reached the new location. Strip back 8 inches of sheathing from the new cable, thread the wires through the hook of the fish tape, then tape the connection. Pull the fish tape back through the existing box, bringing the cable with it *(left, bottom)* and leaving 8 inches of cable at the new location to make the wire connections. To protect the cable, cover the notches in the studs with metal plates. Once the cable has been pulled through the existing box, connect the cable to the old switch or outlet *(page 118)*. Install the new box *(page 115)* and connect the new outlet or switch *(page 118)*. Patch the holes at the studs with scrap drywall and joint compound.

RUNNING CABLE THROUGH THE CEILING

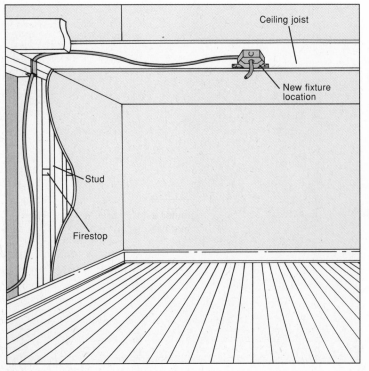

Ceiling joist

New fixture location

Stud

Firestop

1 Planning the run. When adding a ceiling fixture, plan to run cable up a wall and between the ceiling joists to the new location. Turn off power to the circuit you are extending by removing the fuse or tripping the circuit breaker *(page 18)*. Then perform a voltage test at the existing box to confirm that the power is off. Pull the outlet or switch from the wall and disconnect it. Consult the chart on page 118 to determine whether the box is large enough to accommodate additional wires. Mark the location of the ceiling box and, wearing protective goggles, cut its hole. Shine a flashlight into the hole to determine the direction of the joists; you will have to fish the cable up a wall that allows access to the space between the joists. Measure the amount of cable you need, then add 20 percent. You will need two fish tapes and a helper to run the cable.

RUNNING CABLE THROUGH THE CEILING (continued)

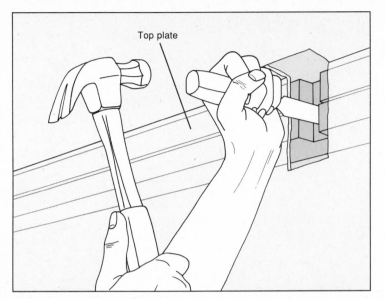

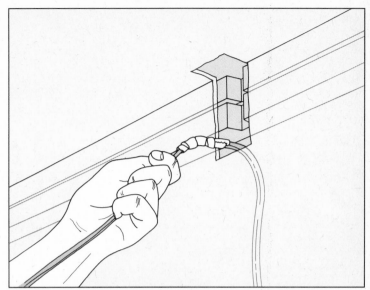

2 **Notching the top plate.** Before feeding the new cable into the existing box, you must make a corner notch in the top plate where the wall meets the ceiling. Drill 3/8-inch pilot holes, then cut an access hole in the corner with a keyhole saw or jigsaw. Use a wood chisel to carve a notch in the top plate to form a passage for the cable *(above)*.

3 **Fishing up to the ceiling.** Use the techniques described on pages 109 and 110 to fish the new cable horizontally to an access hole below the corner notch. Then stand on a ladder to feed a fish tape into the corner notch and down through the wall cavity until it reaches the hole. If a firestop blocks the passage, follow the instructions on page 107, steps 4 and 5. Have a helper use the second fish tape to snag the first tape and pull it out of the access hole. Detach the fish tapes and set the second tape aside. Strip back the new cable, thread the wires through the fish tape, then tape the connection. Pull the fish tape up the wall and through the corner notch, bringing the cable with it *(above)*.

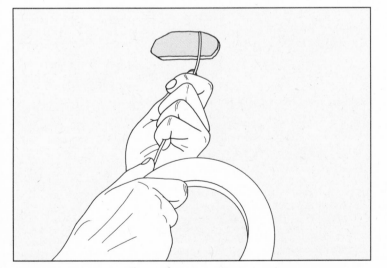

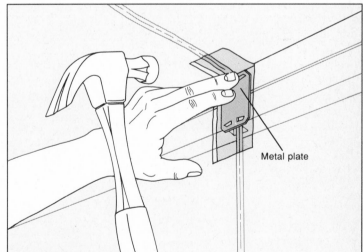

4 **Fishing across the ceiling.** Guide a fish tape from the ceiling fixture hole toward the corner notch, as shown. Have a helper insert a second fish tape (with an opened hook) into the corner notch to snag the first fish tape, then pull the fish tape out of the notch. Detach the second fish tape and attach the cable to the first tape as in step 3. At the new location, pull the fish tape back through the hole in the ceiling, bringing the cable with it. Cut the cable *(page 137)*, allowing 8 inches to make the wire connections.

5 **Protecting the cable.** Hammer a 1/16-inch metal plate over the notch, as shown, to protect the cable. Install the new ceiling box *(page 117)* and connect the fixture *(page 118)*. Then go to the existing box and connect the cable to the old switch or outlet *(page 118)*. Finally, patch the holes you made at the top plate with scrap drywall and joint compound.

INSTALLING RACEWAY

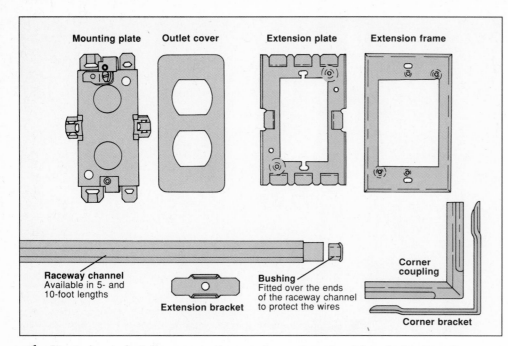

Mounting plate   Outlet cover   Extension plate   Extension frame

Raceway channel
Available in 5- and
10-foot lengths

Extension bracket

Bushing
Fitted over the ends
of the raceway channel
to protect the wires

**Corner
coupling**

Corner bracket

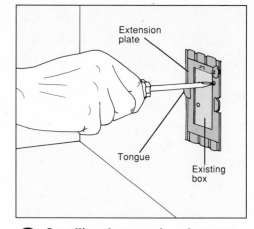

Extension
plate

Tongue

Existing
box

1 **Preparing to install raceway.** Raceway is a system of metal or plastic channels
mounted on a wall to protect the cables or wires that travel through them. Raceway
offers a safe and easy way to extend a circuit, but because it is not concealed, it is best
suited for garages, basements or workshops. First turn off the power to the circuit you are
extending by removing the fuse or tripping the circuit breaker *(page 18)*. Then perform a voltage test at the existing outlet to confirm that the power is off. Pull the outlet from the box and
disconnect it. Mark the location of the new box *(page 115)* and cut its hole, then measure the
amount of raceway you will need. Buy the raceway channel *(above)* and cable or wires,
depending on the size of the raceway channel. You will need an extension plate and frame
for the existing outlet, and a mounting plate and raceway cover for the new outlet. You may
also need extension brackets for long runs or, as in this case, a corner bracket.

2 **Installing the extension plate.** At the
old outlet, position the extension plate
squarely over the existing box and
screw it onto the box, as shown, then pull
the wires out of the box through the extension
plate. Draw a straight and level line from the
tongue on the side of the extension plate
to the hole at the new location.

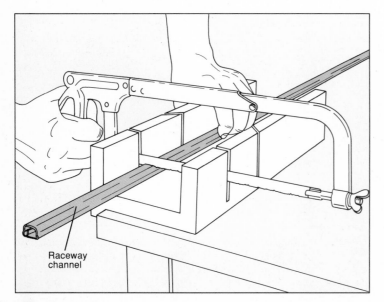

Raceway
channel

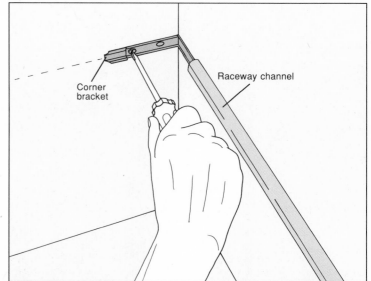

Corner
bracket

Raceway channel

3 **Cutting the raceway.** When running raceway around a corner,
cut the channel to fit the distance from the extension plate at the
existing box to the corner bracket, then from the corner bracket
to the new location. Subtract 3 inches for the corner coupling. Use
a miter box and hacksaw to ensure accurate cuts *(above)*, then file
down any burrs or rough edges.

4 **Mounting the raceway.** Fit one end of the raceway channel
into the extension plate at the existing box and slip the other end
into the corner bracket. Next, place the corner bracket along the
straight line and screw it into the wall, as shown. To protect the wires
from damage, push a bushing into the raceway channel at both ends.

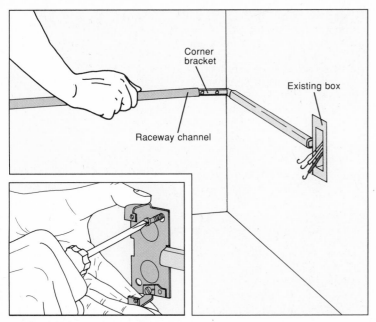

5 **Extending the raceway to the new location.** Fit one end of the second length of raceway channel into the corner bracket, as shown, then slip the other end of the channel into the mounting plate at the new location. Twist off all the other tongues from the mounting plate. Attach the mounting plate to the wall, using screws or drywall anchors to secure it *(inset)*.

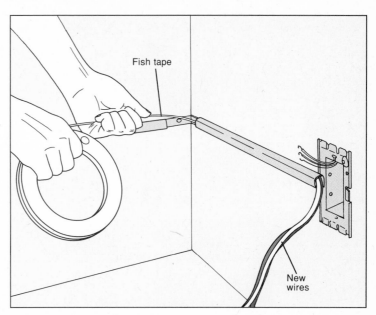

6 **Fishing the new wires.** Feed a fish tape into the raceway channel at the corner and push it through to the existing box. To attach the wires to the fish tape, thread them through the hook, then tape the connection. Pull the fish tape back through the corner bracket, as shown, bringing the wires with it, then detach the fish tape. At the new location, feed the fish tape into the raceway channel toward the corner bracket. Reattach the wires to the fish tape, then pull the tape and wires to the new location.

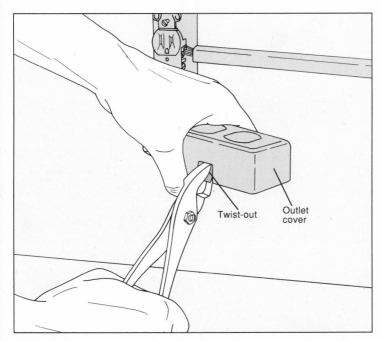

7 **Hooking up the new outlet.** Connect the new outlet *(page 118)*, then screw it onto the mounting plate, making sure that the outlet is straight. Use pliers to pull off the twist-out on the side of the outlet cover, as shown. Snap on the outlet cover and secure it by tightening the mounting screw.

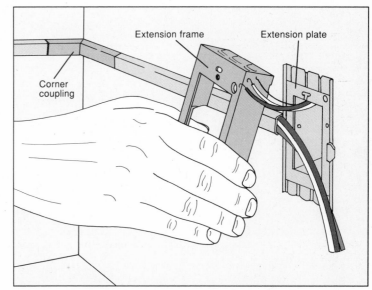

8 **Completing the installation.** Snap the corner coupling onto the corner bracket, taking care not to pinch the wires. Slip the extension frame onto the plate *(above)* and pull the wires through the frame. Reconnect the old outlet *(page 118)*, then screw the outlet to the frame. Finally, screw on the cover plate.

ELECTRICAL BOXES

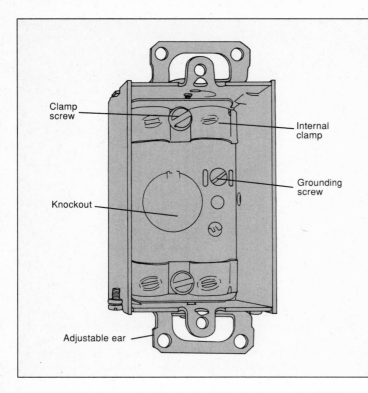

Clamp screw

Internal clamp

Grounding screw

Knockout

Adjustable ear

Anatomy of a box. All electrical connections must be enclosed in an electrical box with a removable cover that permits inspection and repair. Electrical boxes vary in shape, depending on their location and function. Wall boxes are rectangular, ceiling boxes are octagonal or circular and junction boxes are octagonal or square. Box sizes vary according to the number of wires they house *(page 118)*; the standard wall box is 2 1/2 inches deep.

Although metal is the most common material for electrical boxes, plastic boxes have been widely used in homes built since 1970. Metal boxes are more appropriate for circuit extensions, however, since they can be mounted in several different ways.

Any cable entering a box must be secured by means of a strain-relief clamp. If a new box does not already have internal clamps, purchase external clamps *(page 115)*. Most boxes also contain knock-outs—metal discs that can be pried or hammered out to allow cable to enter the box. Knockouts are located behind the internal clamps and on the top, bottom or sides of the box.

All metal boxes must be grounded; bare copper grounding wires entering the box are connected to a grounding screw at the back of the box. Some plastic boxes have a grounding screw on a metal bar.

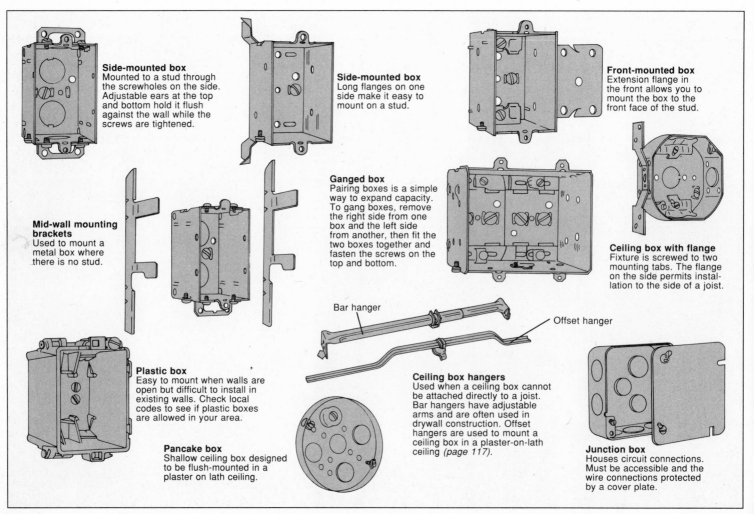

Side-mounted box
Mounted to a stud through the screwholes on the side. Adjustable ears at the top and bottom hold it flush against the wall while the screws are tightened.

Side-mounted box
Long flanges on one side make it easy to mount on a stud.

Front-mounted box
Extension flange in the front allows you to mount the box to the front face of the stud.

Mid-wall mounting brackets
Used to mount a metal box where there is no stud.

Ganged box
Pairing boxes is a simple way to expand capacity. To gang boxes, remove the right side from one box and the left side from another, then fit the two boxes together and fasten the screws on the top and bottom.

Ceiling box with flange
Fixture is screwed to two mounting tabs. The flange on the side permits installation to the side of a joist.

Plastic box
Easy to mount when walls are open but difficult to install in existing walls. Check local codes to see if plastic boxes are allowed in your area.

Pancake box
Shallow ceiling box designed to be flush-mounted in a plaster on lath ceiling.

Bar hanger

Offset hanger

Ceiling box hangers
Used when a ceiling box cannot be attached directly to a joist. Bar hangers have adjustable arms and are often used in drywall construction. Offset hangers are used to mount a ceiling box in a plaster-on-lath ceiling *(page 117)*.

Junction box
Houses circuit connections. Must be accessible and the wire connections protected by a cover plate.

INSTALLING A WALL BOX IN DRYWALL

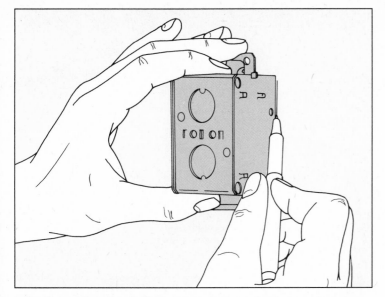

1 **Preparing the wall.** Select the location for the new box, then use a stud finder *(page 132)* or drill small pilot holes to locate a nearby stud. If the stud is too far from the spot you have chosen, you will have to mount the box mid-wall. Set the box squarely against the wall and trace its outline, as shown. If the box has ears at the top and bottom, do not trace or cut them out. Drill a 3/8-inch pilot hole through the wall inside each corner of the outline, then carefully cut along the pattern with a keyhole saw or a jigsaw. Using the fishing techniques described on pages 105-110, run cable to the new location, then continue to step 2.

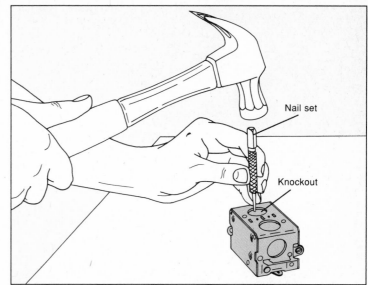

2 **Preparing the box.** If the box has internal clamps, unscrew one of them, remove a knockout behind the clamp and go to step 3. If the box does not have internal clamps, remove one of the knockouts in the side, top or bottom. Punch out the disc with a nail set and hammer, as shown, or pry it off with a screwdriver.

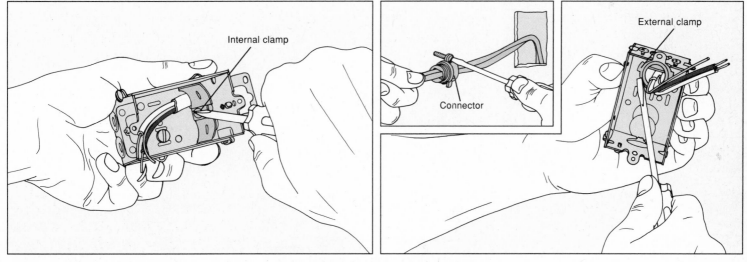

3 **Securing the cable.** To secure the cable to the box, pull the cable through the knockout behind the internal clamp. Reposition the clamp and tighten it so that it grips the sheathing firmly but does not gouge the cable *(above, left)*. External clamps come in two parts. Before inserting the cable into the box, slip the connector onto the cable, then use a screwdriver to fasten it firmly to the cable sheathing just above the stripped wires *(inset)*. Thread the wire ends into the box through the knockout hole and push the connector into the hole. Slip the lock-nut over the wire ends from inside the box, and screw it onto the connector *(above, right)*. If you are mounting the box to a stud, go to the next step. If you are mounting it mid-wall, go to step 5.

INSTALLING A WALL BOX IN DRYWALL (continued)

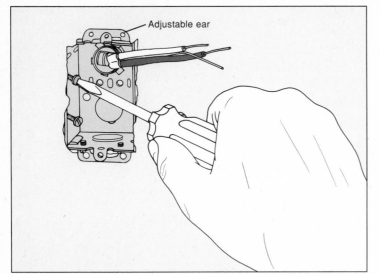

Adjustable ear

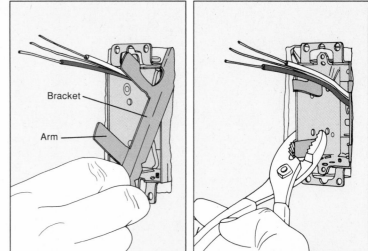

Bracket

Arm

4 **Mounting the box to a stud.** To install a standard side-mounted box, slide forward the adjustable ears on the top and bottom of the box so that they are flush with the edge of the box, then tighten the screws to secure them. Set the box into the hole so that it is flush with the wall, then screw it to the stud, as shown. Connect the new outlet or switch *(page 118)*.

5 **Mounting the box mid-wall.** Where there is no stud, use mid-wall mounting brackets to secure the box to the wall. Slide forward the adjustable ears on the top and bottom of the box so that they are flush with the edge of the box, then tighten the screws. Set the box into the hole. Insert a bracket into the space between the wall and the box *(above, left)*, sliding it up until you can slip the lower part into the wall. Holding the bracket by its arms, pull it forward as far as possible, then bend the arms over the edge of the box, pinching them tightly against the side of the box to avoid later contact with live parts. Insert the second bracket on the other side of the box in the same way *(above, right)*. Connect the new outlet or switch *(page 118)*.

INSTALLING A WALL BOX IN PLASTER ON WOOD LATH

Wood lath

Keyhole saw

Making a hole in a plaster wall. If your house is more than 30 years old, the plaster walls will probably be laid over wood lath. To make an opening for the box, chisel a hole large enough to expose one lath. Mark the midpoint in the exposed lath, then draw a horizontal line from the midpoint along the wall.

You will need a box with adjustable ears. Position the box so that its center aligns with the mark on the wall, then trace the outline of the box (but not the curve of the ears). Use a utility knife to score the outline, then chisel the plaster away to expose the wood lath. Cut the lath with a keyhole saw, as shown. Next, chisel away enough plaster at the top and bottom of the hole to accommodate the ears on the box. Make a knockout hole in the box and secure the cable to it, as on page 115, steps 2 and 3. Set the box in the wall so that it is flush with the plaster surface, then screw the ears to the wood lath.

INSTALLING A CEILING BOX IN DRYWALL

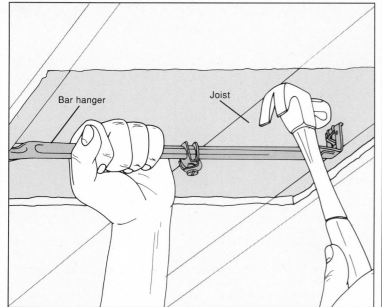

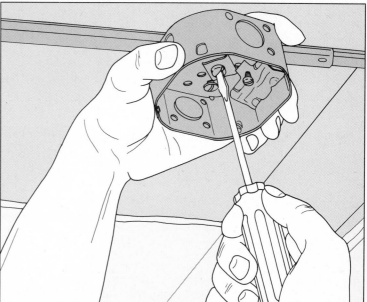

Securing the box with a bar hanger. A ceiling box can be mounted in one of three ways. It can be attached directly to a joist (much like a wall box), it can be suspended between two joists by means of a bar hanger *(above, left)* or it can be hung from the bottom of joists by an offset hanger *(below)*. To install a bar hanger in drywall, first find the joists. Shine a flashlight into the hole you have fished through on page 110, then use a ruler or tape measure to establish the exact location of the joists. Mark the location on the ceiling, then draw a rectangle that stretches from joist to joist. Wearing protective goggles, drill a 3/8-inch pilot hole inside each corner of the outline, then cut along the pattern with a keyhole saw. Use a chisel to remove 1 inch of drywall along the bottom edge of both joists; you will attach the new drywall to this surface when patching the hole. Hammer the ends of the bar hanger to the joists *(above, left)*, then secure the ends with screws.

Pry off or hammer out two knockouts in the ceiling box; you will mount the box through the center knockout and pass the cable through a knockout behind the internal clamp. Screw the box to the hanger *(above, right)*. Pull 8 inches of cable through the knockout hole behind the internal clamp and tighten the clamp, being careful not to gouge the sheathing. With a utility knife, cut a piece of drywall to fit the hole in the ceiling. Coat the edge of the ceiling box with lipstick or colored chalk and position the patch in the ceiling, pressing it against the edge of the box. Then follow the lipstick outline to cut an opening for the box in the patch. Use drywall screws to fasten the patch along the edge of the joists. Using a flexible putty knife, lay a thin coat of joint compound on the seams, then embed joint tape into the wet compound. Cover the tape with another thin coat. When the compound is dry, sand it smooth. Connect the new fixture *(page 118)*.

INSTALLING A CEILING BOX IN PLASTER ON WOOD LATH

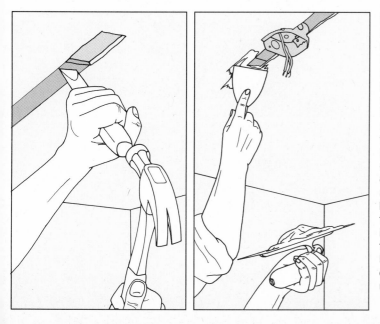

Securing the box with an offset hanger.
Mark the spot where you plan to locate the ceiling box. Wearing protective goggles, use a hammer and chisel to remove the plaster at that location. Chisel a 2-inch-wide channel along the lath *(far left)*, until you expose the nails where the lath is joined to the joist. Extend the channel in the opposite direction until you reach the other joist. Trace the box on the ceiling and remove the plaster from the lath, then use a keyhole saw to cut out the exposed lath. Run the cable from the existing outlet to the hole *(page 110)*. Pry off or hammer out the center knockout and a knockout behind the internal clamp. Mount the box on the hanger. Pull 8 inches of cable through the internal clamp and tighten it without gouging the sheathing. Fasten each end of the offset hanger to the joists with screws. Patch the channels with joint compound *(near left)*. Connect the new fixture *(page 118)*.

CONNECTING A NEW OUTLET, SWITCH OR FIXTURE

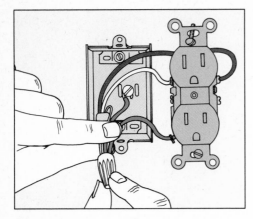

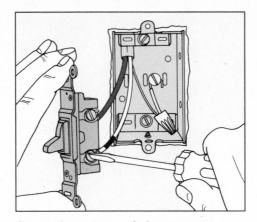

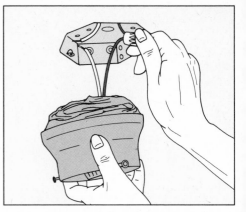

Connecting a new outlet. Strip the insulation from the white and black wires and shape each wire end into a hook *(page 138).* Wrap the black wire around a brass terminal and the white wire around a silver terminal and tighten the connections. Attach a jumper wire to the grounding screw at the back of the box and another jumper to the green grounding terminal on the outlet, then twist the wires together with the bare grounding wire from the cable and screw on a wire cap.

Connecting a new switch. Strip the insulation from the white and black wires and shape each wire end into a hook *(page 138).* Attach the black wire to one terminal screw and tighten the connection. Wrap black electrical tape around the white wire to identify it as hot, then connect it to the other switch terminal and tighten the connections *(above).* Attach the bare grounding wire to the grounding screw at the back of the box.

Connecting a new fixture. Strip the insulation from the white and black wires *(page 138).* Twist the black wire end together with the black fixture lead, and the white wire end together with the white fixture lead, and screw a wire cap onto each connection *(above).* Attach the bare grounding wire to the grounding screw in the ceiling box.

EXTENDING A CIRCUIT FROM AN EXISTING BOX

Electrical boxes are designed to hold a specific number of wires. To determine whether the existing box is large enough to accommodate extra wires, measure the box, then count the existing wires and add the number of new wires. Grounding wires count as one wire; if there are three grounding wires in a box, they still count as just one wire. Jumper wires and external clamps are not counted. Any internal clamps, studs or nipples in the box are counted as one wire; outlets or switches are counted as another. Refer to the chart at right. If the box is not large enough to hold the new connections, expand it *(page 114),* install a deeper box or choose another location.

MAXIMUM WIRES IN A BOX

	Wall Box Size			Ceiling Box Size		
	2 1/2"	2 3/4"	3 1/2"	1 1/4"	1 1/2"	2 1/8"
No. 10 wire	5	5	7	5	6	8
No. 12 wire	5	6	8	5	6	9
No. 14 wire	6	7	9	6	7	10

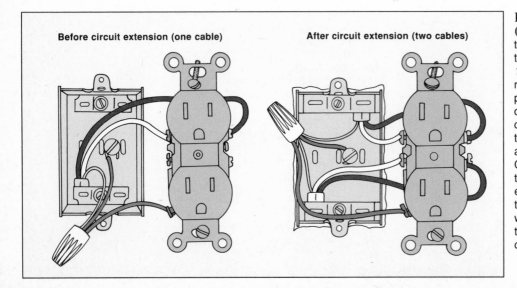

Before circuit extension (one cable)

After circuit extension (two cables)

Extending from an end-of-the-run outlet (one cable enters the box). Turn off power to the circuit by removing the fuse or tripping the circuit breaker at the service panel *(page 18).* To confirm that the power is off and remove the outlet, follow the instructions on page 86. Use one of the techniques described on pages 105-110 to run a second cable into the box. Strip the insulation from the wires of the new cable and curl the white and black wire ends into a hook *(page 138).* Connect each of the black wires to a brass terminal on one side of the outlet. Connect each of the white wires to a silver terminal on the opposite side of the outlet. Unscrew the wire cap from the grounding wire connection, twist in the bare grounding wire from the new cable and screw the wire cap back on.

EXTENDING A CIRCUIT FROM AN EXISTING BOX (continued)

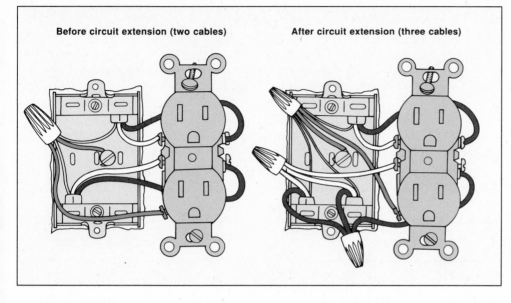

Before circuit extension (two cables)   After circuit extension (three cables)

Extending from a middle-of-the-run outlet (two cables enter the box). Turn off power to the outlet *(page 18)*. To confirm that the power is off and remove the outlet, follow the instructions on page 86. Use one of the techniques described on pages 105-110 to run a third cable into the box. Connect a black wire end to a brass outlet terminal. Twist the other two black wires together with a jumper, screw on a wire cap and attach the jumper to the other brass terminal. Connect a white wire end to a silver outlet terminal. Twist the other two white wires together with a jumper, screw on a wire cap and attach the jumper to the other silver terminal. Unscrew the grounding connection, twist in the bare grounding wire from the new cable and screw the wire cap back on. Attach the grounding jumper to the green grounding terminal on the outlet.

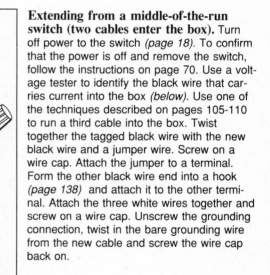

Before circuit extension (two cables)   After circuit extension (three cables)

Extending from a middle-of-the-run switch (two cables enter the box). Turn off power to the switch *(page 18)*. To confirm that the power is off and remove the switch, follow the instructions on page 70. Use a voltage tester to identify the black wire that carries current into the box *(below)*. Use one of the techniques described on pages 105-110 to run a third cable into the box. Twist together the tagged black wire with the new black wire and a jumper wire. Screw on a wire cap. Attach the jumper to a terminal. Form the other black wire end into a hook *(page 138)* and attach it to the other terminal. Attach the three white wires together and screw on a wire cap. Unscrew the grounding connection, twist in the bare grounding wire from the new cable and screw the wire cap back on.

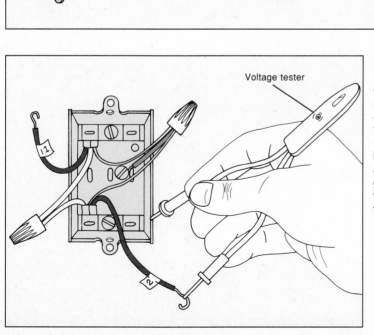

Voltage tester

Identifying the incoming hot black wire. Use a voltage tester to identify the black wire that carries the current into the box. With power confirmed off, separate the black wires and make sure that they are not touching each other or the metal box. Wrap a piece of masking tape around each black wire and give each a number. Then turn on the power to the circuit. (**Caution:** Do not touch any wire ends or the metal box. Use the one-hand technique shown or be sure to hold the tester probes by their insulated handles.) Touch one probe to the grounded metal box and the other to each black wire end in turn, as shown. The tester will glow when the probe touches the hot (current-carrying) black wire. Note the number of this wire, then turn off the power and remove the masking tape from the other wire.

OUTDOOR LIGHTING

Outdoor lighting has both a practical and an aesthetic side. Post lights illuminate sidewalks, driveways and garages; porch lights welcome visitors and discourage intruders; and patio lights can make a large property seem intimate or a small yard spacious.

Outdoor fixtures may be permanently wired to a house circuit—a porch light mounted beside a door, for example—or temporarily plugged into an outdoor outlet, such as a string of patio lanterns. In either case, they must be protected from the elements and accidental grounding by means of waterproof covers and insulation. The bulb, too, should be weatherproof to prevent its shattering from contact with rain or snow.

Depending on local codes, outdoor fixtures are served by heavy, sheathed cable buried in the lawn, or by wiring fed through plastic or metal conduit. The wiring usually taps into an indoor circuit at a porch light or through the foundation or eave of the house. Sometimes there is an outlet or switch, encased in an exterior-grade box, installed along the circuit.

You can lengthen the life of your outdoor lighting system significantly with regular maintenance. Periodically remove

TROUBLESHOOTING GUIDE

SYMPTOM	POSSIBLE CAUSE	PROCEDURE
PORCH LIGHTS		
Bulb flickers or does not light	No power to the circuit	Check for blown fuse or tripped circuit breaker *(p. 18)* □○
	Bulb loose or burned out	Tighten or replace bulb
	Indoor wall switch faulty	Check switch *(p. 70)*
	Socket contact dirty or bent too far down	Clean or bend up tab *(p. 122)* □○
	Socket connections loose or dirty	Clean and tighten connections *(p. 122)* □○
	Socket and wires faulty	Test and replace socket and wires *(p. 122)* ◨●
POST LIGHTS		
Bulb flickers or does not light	No power to the circuit	Check for blown fuse or tripped circuit breaker *(p. 18)* □○
	Bulb loose or burned out	Tighten or replace bulb
	Indoor wall switch faulty	Check switch *(p. 70)*
	Socket contact dirty or bent too far down	Clean or bend up tab *(p. 124)* □○
	Socket connections loose or dirty	Clean and tighten connections *(p. 124)* □○
	Socket and wires faulty	Test and replace socket and wires *(p. 124)* ◨●
	Outdoor circuit faulty	Have an electrician run a new circuit
PLUG-IN PATIO LIGHTS		
Bulb flickers or does not light	Plug not firmly in outlet	Push plug in firmly
	Bulb loose or burned out	Tighten or replace bulb
	Indoor wall switch faulty	Check switch *(p. 70)*
	Socket contact dirty or bent down too far	Clean or bend up tab *(p. 126)* □○
	Socket and wires faulty	Test and replace socket and wires *(p. 126)* ◨●
	Outlet connections loose or dirty	Clean and tighten connections *(p. 126)* □○
	Outlet faulty	Replace outlet *(p. 127)* ◨●
	Plug or cord damaged	Replace plug *(p. 128)* □○ or plug and cord *(p. 128)* ◨●
LOW-VOLTAGE PATIO LIGHTS		
One light not working	Bulb loose or burned out	Tighten or replace bulb
	Socket and wires faulty	Test and replace socket and wires *(p. 129)* ◨●
	Wiring between lights faulty	Replace wiring *(p. 129)* □○
More than one light not working	Transformer connections loose or dirty	Clean and tighten connections *(p. 130)* □○
	Transformer faulty	Test and replace fuse *(p. 130)* □○ or transformer *(p. 130)* ◨●▲

DEGREE OF DIFFICULTY: □ Easy ◨ Moderate ■ Complex
ESTIMATED TIME: ○ Less than 1 hour ◑ 1 to 3 hours ● Over 3 hours

▲ Multitester required

dirt, leaves and other debris from the fixtures. After making sure that they have cooled, wipe both the bulb and fixture with a damp cloth to remove dust and insects. Replace a burned-out standard bulb with a weatherproof bulb or a long-life bulb that requires less frequent replacement *(page 28)*.

When repairing any part of an outdoor lighting system, seal all connections by wrapping the wire caps with electrical tape. At the same time, inspect each housing and gasket for signs of damage. If any part is cracked, corroded or broken, replace it before the wiring inside is affected. The joint between the

fixture and the wall should also be made watertight with a caulking compound.

For maximum safety, turn off power to the circuit, even to clean or change a light bulb. Stand on a wooden plank or rubber mat when working on damp ground and use a wooden ladder to get at hard-to-reach fixtures. Do not touch any terminals, bare wire ends or the metal box until you have confirmed that the power is off with a voltage tester. Unplug a plug-in patio light and, when working with a low-voltage system, unplug the transformer from the 120-volt household current.

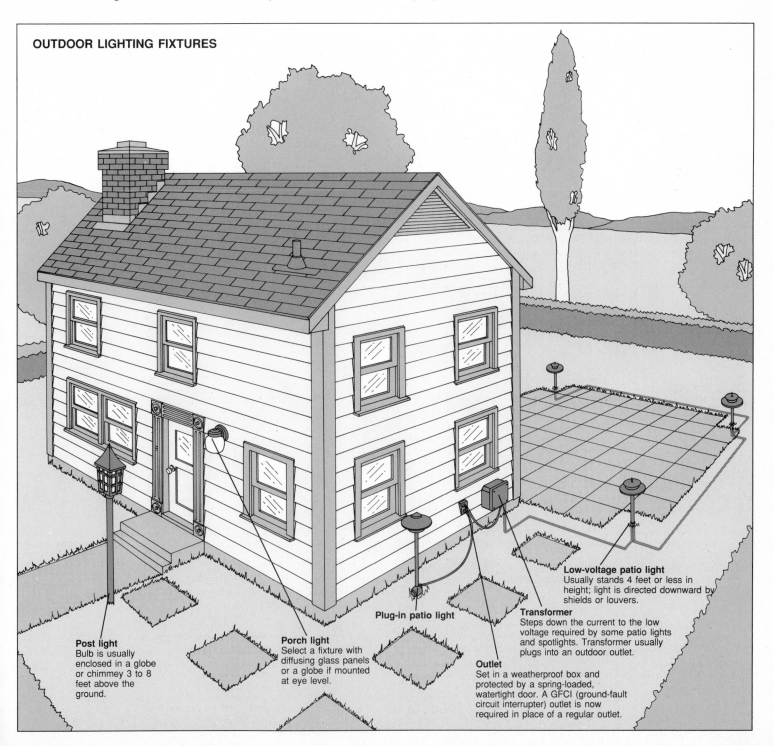

OUTDOOR LIGHTING FIXTURES

Post light
Bulb is usually enclosed in a globe or chimney 3 to 8 feet above the ground.

Porch light
Select a fixture with diffusing glass panels or a globe if mounted at eye level.

Plug-in patio light

Outlet
Set in a weatherproof box and protected by a spring-loaded, watertight door. A GFCI (ground-fault circuit interrupter) outlet is now required in place of a regular outlet.

Transformer
Steps down the current to the low voltage required by some patio lights and spotlights. Transformer usually plugs into an outdoor outlet.

Low-voltage patio light
Usually stands 4 feet or less in height; light is directed downward by shields or louvers.

SERVICING PORCH LIGHTS

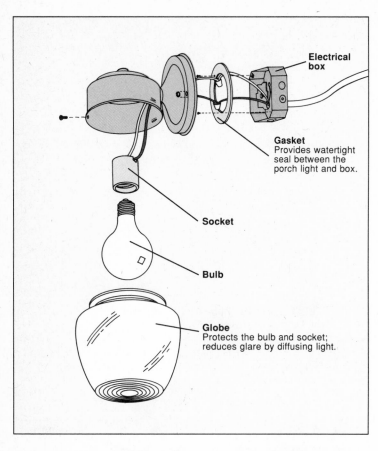

Electrical box

Gasket
Provides watertight seal between the porch light and box.

Socket

Bulb

Globe
Protects the bulb and socket; reduces glare by diffusing light.

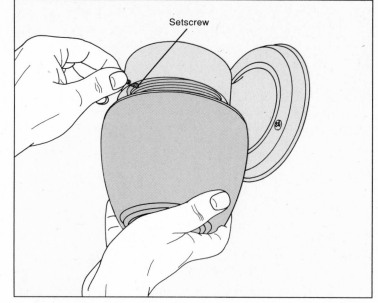

Setscrew

1 **Removing the globe.** Turn off power to the porch light by removing the fuse or tripping the circuit breaker *(page 18)*. Support the globe with one hand and loosen any screws holding it in place, as shown, then lower the globe. If the bulb is loose, tighten it. Unscrew a burned-out bulb and put in a replacement of the same wattage. Turn on the power and flip on the switch. If the fixture lights, reinstall the globe. If it doesn't, go to the next step.

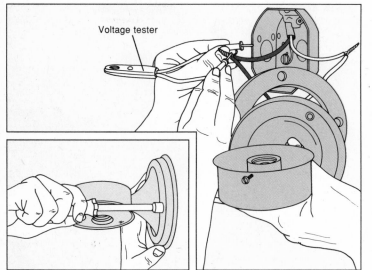

Voltage tester

2 **Testing for voltage.** Flip off the wall switch and turn off the power. Loosen the screws holding the fixture to the wall *(inset)* and pull it slightly forward. Have a helper hold the fixture while you pull the wires out of the box, then remove the wire caps to expose the connections, taking care not to touch the wire ends. Use a voltage tester to confirm that the power is off by touching one of the tester probes to the black wire connection and the other probe first to the grounded metal box, as shown, then to the white wire connection. Then test between the white wires and the grounded metal box. The tester should not glow in any test. If it does, return to the service panel and turn off power to the correct circuit. When the power is confirmed off, untwist the connections and take down the fixture.

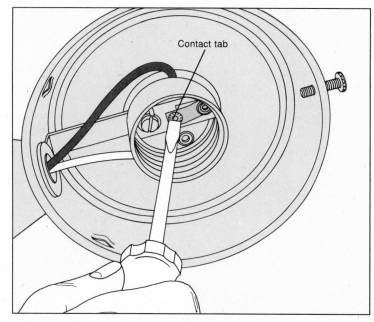

Contact tab

3 **Servicing the socket.** Inspect the socket for wear or corrosion. Use an old knife or the blade of a screwdriver to scrape clean the contact tab, as shown, and bend up the tab slightly to improve contact with the bulb.

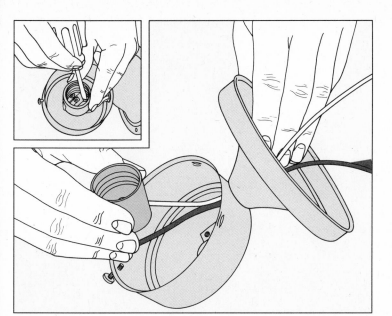

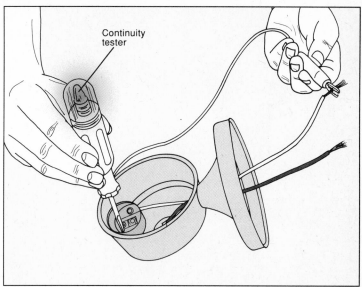

4 **Servicing the socket connections.** Loosen the mounting screw that holds the socket in place *(inset)*, then pull the socket far enough out of the fixture to expose the wire connections *(above)*. If the socket has preattached wires, go to the next step. If the wires are connected to terminal screws, clean or tighten the connections, or clip back damaged wire ends and reconnect them *(page 141)*.

5 **Testing the socket.** Place the alligator clip of a continuity tester on the white wire end and touch the tester probe to the threaded metal tube of the socket, as shown. Then place the alligator clip on the end of the black wire and touch the tester probe to the socket contact tab. The tester should light in both tests. If the socket passes the test, reinstall it and remount the porch light *(step 7)*. If the tester doesn't light, replace the socket and wires *(next step)*.

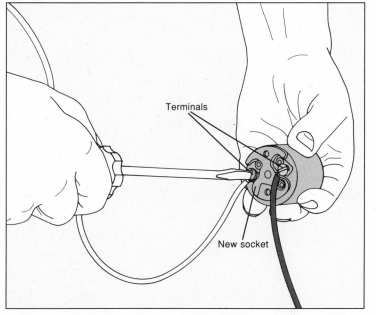

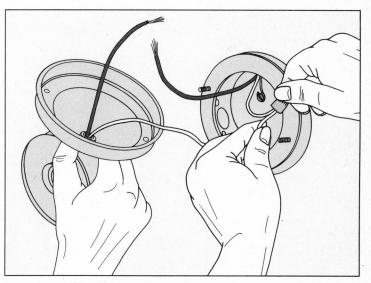

6 **Replacing the socket.** Buy a replacement socket with preattached wires, or a socket with terminal screws and two pieces of wire of the same gauge, length and color as the old ones. To connect the wires to socket terminals, strip back the insulation *(page 138)*, twist the strands together and bend them into a hook, then loop the black wire around the brass terminal and the white wire around the silver terminal. Tighten the connections *(above)*. Install the socket by feeding the wires into the fixture and pulling them through. Tighten the mounting screw to secure the socket to the fixture.

7 **Mounting the porch light.** Strip back the wires inside the box to improve the connections *(page 141)*. Before completing the repair, replace the gasket if it shows signs of wear. Have a helper hold the fixture while you twist the ends of the white wires together and screw on a wire cap, as shown. Repeat the procedure for the other set of wires. To ensure a watertight connection, wrap the wire caps with electrical tape. Gently fold the wires back into the box. Set the fixture against the wall and tighten the mounting screws. Screw in a light bulb, then turn on the power and flip on the wall switch to test the fixture. Then use a caulking gun to apply a thin line of exterior-grade silicone sealant around the outside rim of the base. Replace the globe and tighten any screws that hold it in place.

SERVICING POST LIGHTS

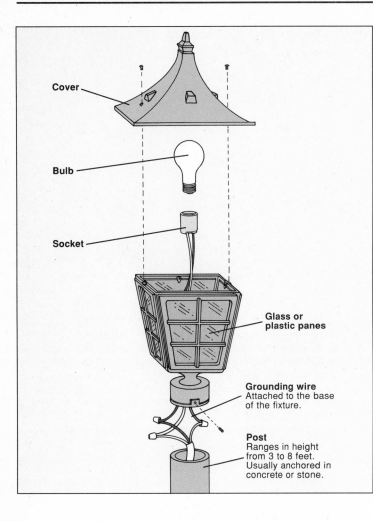

Cover

Bulb

Socket

Glass or plastic panes

Grounding wire
Attached to the base of the fixture.

Post
Ranges in height from 3 to 8 feet. Usually anchored in concrete or stone.

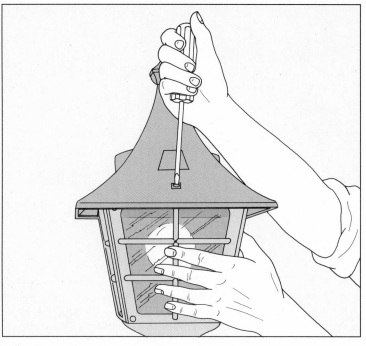

1 **Removing the cover.** Turn off power to the fixture by removing the fuse or tripping the circuit breaker *(page 18)*. Loosen any screws holding the cover in place, as shown, and lift it free of the fixture. If the bulb is loose, tighten it. Unscrew a burned-out bulb and screw in a replacement of the same wattage. Then turn on the power and flip on the switch. If the bulb still doesn't light, turn off the power again and continue to the next step.

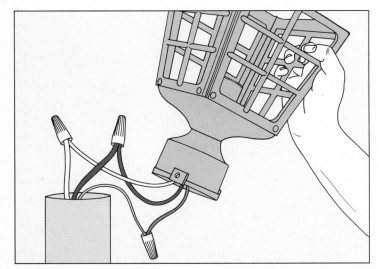

2 **Access to the wiring.** If the post light has glass or plastic panes in the sides, remove them before working on the fixture. Loosen any screws securing the fixture to the post. Lift the fixture off the post, pulling the wire connections clear *(above)*. Have a helper hold the fixture while you unscrew the wire caps, taking care not to touch any exposed wire ends.

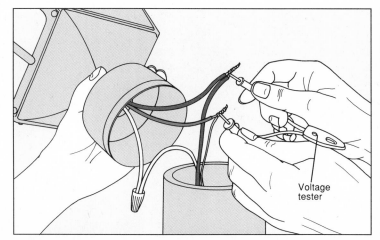

Voltage tester

3 **Testing for voltage.** Use a voltage tester to confirm that the power is off by touching one of the tester probes to the black wire connection and the other probe first to the grounding wire connection, as shown, then to the white wire connection. Then test between the white wires and the grounding wires. The tester should not glow in any test. If it does, return to the service panel and turn off power to the correct circuit. With the power off, separate the wires to detach the fixture. Be sure to hook or tape the incoming wires over the edge of the post to prevent them from slipping down the post.

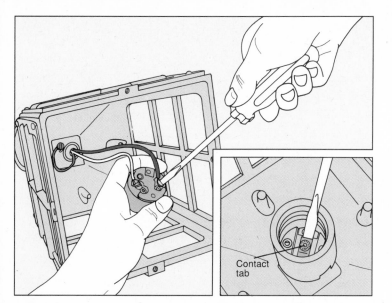

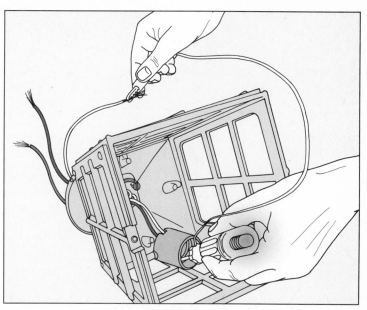

4 **Servicing the socket.** Use an old knife or screwdriver to scrape any corrosion off the socket contact tab, then pry it up slightly *(inset)* to improve contact with the bulb. Loosen the mounting screw that holds the socket in place and pull the socket far enough out of the fixture to expose the wire connections. If the socket has preattached wires, go to the next step. If the wires are connected to terminal screws, clean or tighten them *(above)*, or cut back damaged wire ends and reconnect them *(page 141)*.

5 **Testing the socket.** Place the alligator clip of a continuity tester on the white wire end and touch the tester probe to the threaded metal tube of the socket, as shown. Then place the alligator clip on the black wire end and touch the tester probe to the socket contact tab. The tester should light in both tests. If the socket tests OK, reinstall it, then remount the fixture *(step 7)*. If the tester doesn't light, remove the socket and wires from the fixture and replace them.

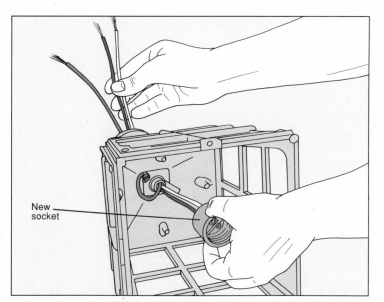

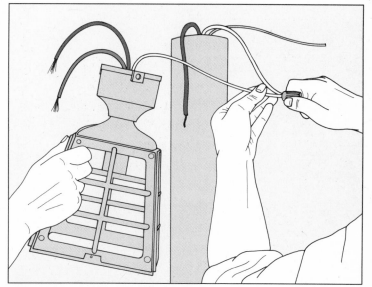

6 **Replacing the socket.** Buy a replacement socket with preattached wires, or a socket with terminal screws and two pieces of wire of the same gauge, length and color as the old ones. To connect the wires to the socket terminals, strip back the insulation *(page 138)*, twist the strands together and bend them into a hook, then loop the black wire around the brass terminal screw and the white wire around the silver screw. Tighten the connections. Install the socket by feeding the wires into the fixture and pulling them through *(above)*. Tighten the mounting screw to secure the socket to the fixture.

7 **Mounting the post light.** Strip back the wires inside the post to improve the connections *(page 141)*. Have someone hold the fixture while you twist the ends of the white wires together and screw on a wire cap, as shown. Repeat the procedure for the black wires, then the grounding wires. To ensure a watertight connection, wrap the wire caps with electrical tape. Gently fold the wires into the post. Position the fixture on the post and tighten the retaining screws. Screw in a light bulb, then turn on the power and test the fixture. Finally, slide in the glass panes, if any, and put on the cover.

SERVICING PLUG-IN PATIO LIGHTS

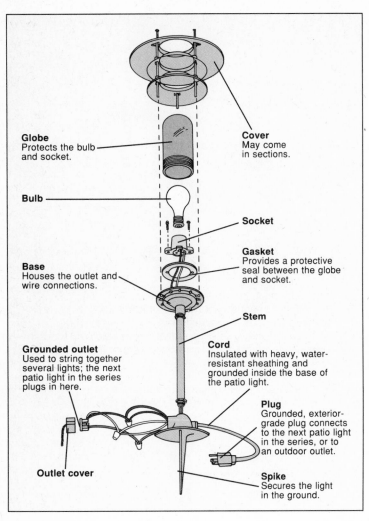

Globe
Protects the bulb and socket.

Cover
May come in sections.

Bulb

Socket

Gasket
Provides a protective seal between the globe and socket.

Base
Houses the outlet and wire connections.

Stem

Cord
Insulated with heavy, water-resistant sheathing and grounded inside the base of the patio light.

Grounded outlet
Used to string together several lights; the next patio light in the series plugs in here.

Plug
Grounded, exterior-grade plug connects to the next patio light in the series, or to an outdoor outlet.

Outlet cover

Spike
Secures the light in the ground.

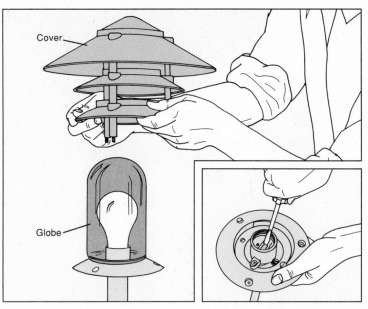

Cover

Globe

1 Servicing the socket. When a plug-in patio light does not operate correctly, first inspect the cord and plug. Replace them if they show signs of damage *(page 128)*. If the cord and plug are in good condition, plug the patio light into another outlet. If it works, check the original outlet *(opposite page)*. If the light still doesn't work, unplug it, loosen the screws holding the cover in place and lift off the cover *(above)*, then unscrew the globe. Tighten a loose bulb, or remove the bulb and inspect the socket contact tab. Use an old knife or a screwdriver to scrape any corrosion off the tab, then pry it up slightly *(inset)* to improve contact with the bulb. Replace a burned-out bulb with one of the same wattage. Plug in the patio light and test it. If it doesn't work, unplug it again, remove the bulb and service the connections in the base of the patio light *(next step)*.

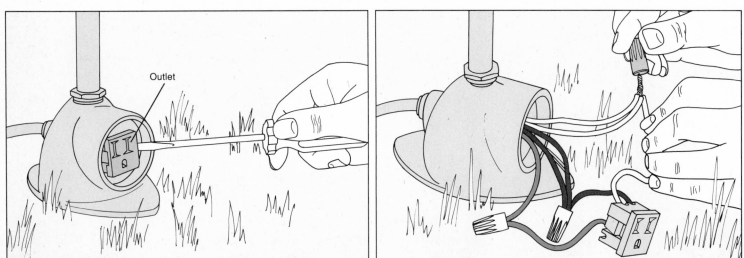

Outlet

2 Servicing the connections. Use a screwdriver to pry the outlet out of the base *(above, left)*. Unscrew the wire caps and separate the connections. If the wire ends are dirty or corroded, clean them with fine sandpaper or, to ensure a good connection, clip the wire ends and strip back the insulation *(page 141)*. Then twist the wire ends together and screw a wire cap on each connection *(above, right)*. Fold the wires into the base of the light, then push in the outlet. Screw in a bulb, plug in the patio light and flip on the switch to test it. If it still doesn't work, test the socket.

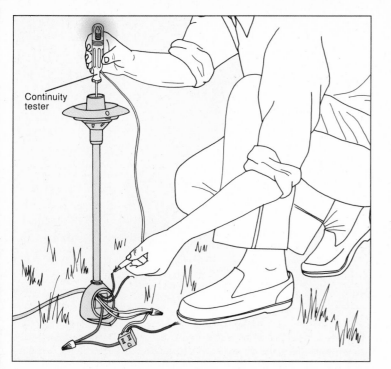

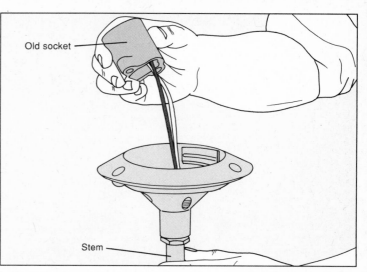

3 **Testing the socket.** Unplug the patio light. Pry the outlet out of the base, then pull out the wires and take apart the connections again. Place the alligator clip of a continuity tester on the end of the black wire leading to the socket and touch the tester probe to the socket contact tab, as shown. Then place the alligator clip on the end of the white wire leading to the socket and touch the tester probe to the threaded metal tube of the socket. The tester should light in both cases. If it doesn't, the socket and its wires should be replaced.

4 **Replacing the socket.** Loosen the screws holding the socket in the patio light, then pull the socket and wires free *(above)*. Buy a replacement socket with preattached wires, or a socket with terminal screws and two pieces of wire of the same gauge, length and color as the old ones. To connect the wires to the socket terminals, strip back the insulation *(page 138)*, then hook the black wire around the brass terminal and the white wire around the silver terminal. Twist together the socket wires, then feed them down the stem until they appear at the base of the patio light. Secure the socket to the patio light. At the base, twist together the black wire ends and screw on a wire cap. Repeat for the white wires, then the grounding wires. Seal the connections by wrapping them with electrical tape, then fold the wires into the base and push the outlet into place. Screw in a bulb and plug in the patio light. If it works, replace the globe and cover. If it doesn't, check the outlet in the next patio light *(below)*.

REPLACING THE OUTLET IN THE NEXT PATIO LIGHT

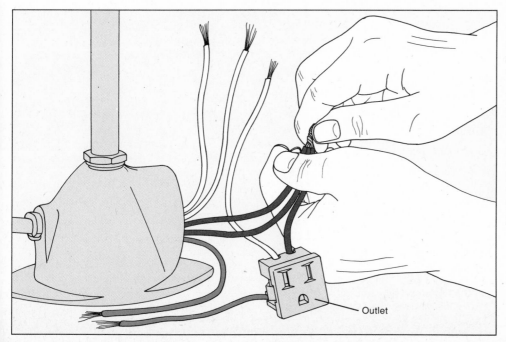

Troubleshooting two types of outlets.
The patio light may be plugged into an outdoor outlet or into another light in the series. If you suspect that the outlet is faulty, service and replace it as you would an indoor outlet *(page 86)*. If the outlet in the base of the next patio light is at fault, unplug that fixture, then pry the outlet out of the base. Detach the wire connections and buy a replacement outlet. Twist together the black outlet lead with the black wire ends *(left)*, the white outlet lead with the white wire ends, and the grounding lead with the grounding wire, then screw on wire caps and seal the connections with electrical tape. Fold the wires into the base, then push in the new outlet.

REPLACING THE PLUG

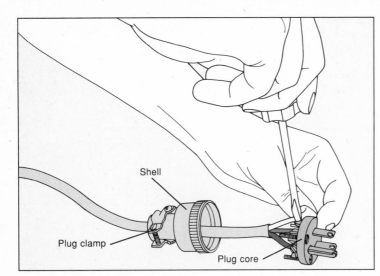

Shell

Plug clamp

Plug core

1 **Connecting the new plug.** Use diagonal-cutting pliers to snip the old plug off the cord. Buy a heavy-duty replacement plug made for outdoor use. Loosen the screws on the face of the plug and slip the plug core out of its rubber shell. Then loosen the plug clamp and slide the shell onto the cord. Strip back 1 1/2 inches of sheathing to expose the three wires, then strip 3/4 inch of insulation from each wire end *(page 138)*. Connect the wire ends to the plug terminals *(above)*, hooking the white wire around the silver terminal, the black wire around the brass terminal and the green grounding wire around the green terminal. Tighten the connections.

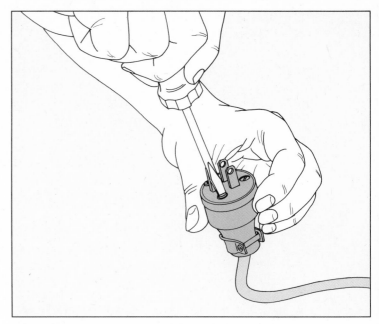

2 **Securing the plug to the cord.** Fit the shell over the plug core, then tighten the screws on the face of the plug *(above)*. Tighten the screws on the plug clamp to secure the plug to the cord.

REPLACING THE CORD

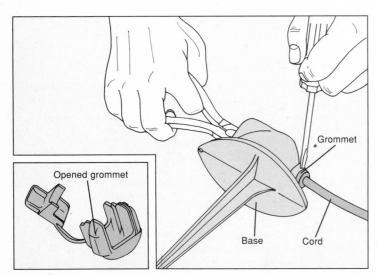

Opened grommet

Grommet

Base   Cord

1 **Removing the old cord.** Unplug the patio light and tug it out of the ground. Pry the outlet from the base of the fixture. Pull the wire connections out of the base and detach them. Unscrew the stem of the patio light and set it aside. Loosen the grounding screw in the base and disconnect the grounding wire. The cord is held in place by a plastic strain-relief grommet. To remove the grommet, grasp it from inside the base with long-nose pliers and push it out, while prying it from the other side with a screwdriver *(above)*. Pull the cord out of the base of the patio light, then pry open the grommet and slip it off the cord. Buy a replacement cord of the same length and gauge as the old one, a new outdoor plug and, if necessary, a new grommet.

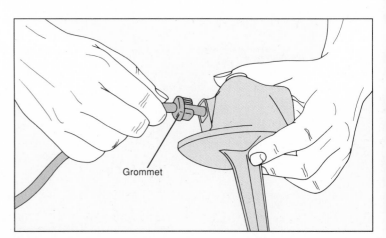

Grommet

2 **Installing the new plug and cord.** Attach the new plug on one end of the new cord *(steps 1 and 2, above)*. Slide the opened grommet on the other end of the cord, strip back about 5 inches of cord sheathing, then strip 3/4 inch of insulation from each wire end *(page 138)*. Snap the grommet closed, securing it to the cord, then feed the wire ends into the base of the patio light *(above)*. Wedge the grommet into the hole in the base. Reassemble the patio light. Twist together the black wires, then the white wires, then the green grounding wires and secure the connections with wire caps and electrical tape. Gently fold the wires into the base and push the outlet back into place. To prevent moisture from seeping into the base, apply a thin bead of silicone sealant around the grommet.

SERVICING LOW-VOLTAGE PATIO LIGHTS

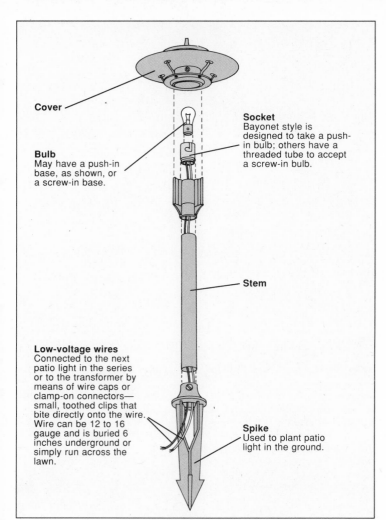

Cover

Bulb
May have a push-in base, as shown, or a screw-in base.

Socket
Bayonet style is designed to take a push-in bulb; others have a threaded tube to accept a screw-in bulb.

Stem

Low-voltage wires
Connected to the next patio light in the series or to the transformer by means of wire caps or clamp-on connectors—small, toothed clips that bite directly onto the wire. Wire can be 12 to 16 gauge and is buried 6 inches underground or simply run across the lawn.

Spike
Used to plant patio light in the ground.

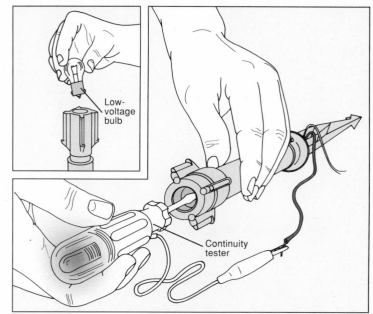

Low-voltage bulb

Continuity tester

1 **Testing the socket.** Unplug the transformer. Unscrew the patio light cover and lift it off. To remove a push-in bulb that has burned out, gently press it down and turn, then lift it out *(inset)*; install a replacement low-voltage bulb. If the new bulb doesn't light, the problem may be in the socket. Tug the patio light out of the ground, then pull the wires from the base of the stem to expose the connections. Unscrew the wire caps and detach the wires. Place the alligator clip of a continuity tester on the black wire end leading from the stem and touch the tester probe to the socket contact *(above)*. Then place the alligator clip on the white wire end and touch the probe to the metal socket tube. The tester should light in both cases. If so, next check the wires between the lights *(step 3)*. If not, replace the socket and wires.

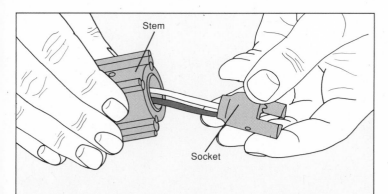

Stem

Socket

2 **Replacing the socket.** Grasp the rim of the socket with long-nose pliers and pull it free of the stem, as shown. For a stubborn socket, separate the head of the patio light from the stem, then use a screwdriver to force up the socket from below. Buy a compatible low-voltage socket with preattached wires. To install the new socket, feed the wires down the stem, then push the socket firmly into place. Before reconnecting the wires at the base of the stem, check for dirty or corroded wire ends and clean or clip them back if necessary *(page 141)*. Then reconnect the wires and reassemble the light, installing the light bulb and cover. Plant the patio light in the ground and turn on the power. If the bulb does not glow, next check the wires between the patio lights.

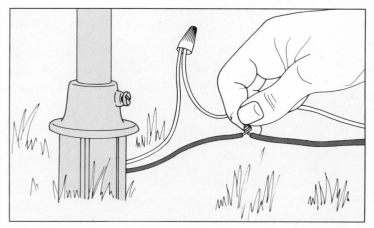

3 **Replacing the low-voltage wires.** Wires exposed to dampness can deteriorate over time. To replace damaged wire, unplug the transformer, tug the patio light a few inches out of the ground and pull the wires out of the base of the stem to take apart the connections. Buy replacement low-voltage wire of the same gauge. Run the new wire along the surface or bury it 6 inches in the ground for better protection. Strip back the wire ends *(page 138)*, then twist each wire end together with a matching wire from the patio light stem *(above)* and screw on wire caps. Plug in the transformer and test the system. If it still doesn't work, check the transformer *(page 130)*.

SERVICING THE TRANSFORMER

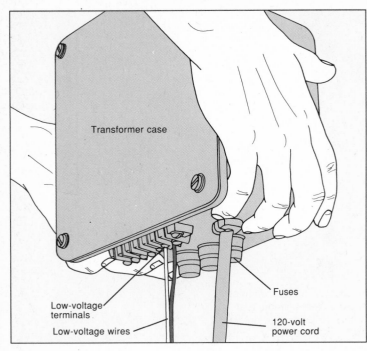

1 **Dismounting the transformer.** The transformer for a low-voltage lighting system is usually housed in a weatherproof case and mounted on a bracket near an outdoor outlet. Before servicing it, unplug it from the outlet. Then lift the case off its bracket *(above)* and set it on a flat, dry surface with the fuses facing up. Inspect the cord and plug; if they show signs of damage, replace the transformer.

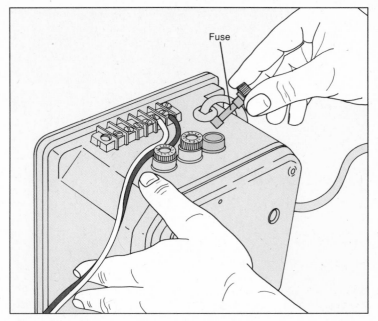

2 **Checking and replacing the fuses.** To remove a fuse, press it in slightly and give it a quarter-turn until it snaps out *(above)*. Replace the fuse if the inner strip is broken or the glass stem is clouded. If there are no signs of damage, test the fuse by touching the alligator clip of a continuity tester to one of the metal ends and the tester probe to the other end. The tester should light. If it doesn't, install an exact replacement. If the fuse passes the test, put it back and inspect and test the other fuses. If they all pass the test, service the low-voltage terminal connections.

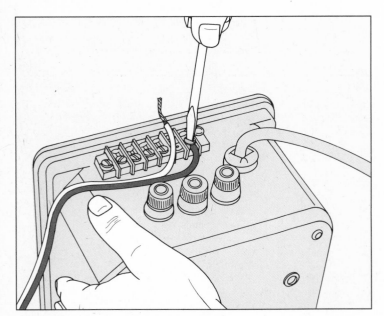

3 **Servicing the terminal connections.** Loosen one terminal screw and unhook its wire. Check the terminal for dirt and corrosion and clean it with fine sandpaper. Cut back the wire and strip the insulation to make a clean connection *(page 141)*, then reattach the wire end to the terminal *(above)*. Repeat the procedure for each connection. Plug in the transformer and flip on the switch. If the system still doesn't work, leave the transformer plugged in and test it *(step 4)*.

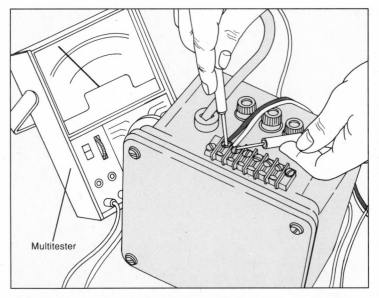

4 **Testing and replacing the transformer.** Use a multitester *(page 135)* to confirm that low-voltage power is being delivered to the lighting system. Set the tester on the ACV scale and turn the dial to the 50-volt range. Taking care to hold the probes by their insulated handles, place them on the low-voltage terminal connections, as shown. If the tester registers little or no voltage, unplug the transformer, disconnect the wires and buy a compatible replacement. Connect the new transformer by hooking the low-voltage wires around the correct terminals, then plug it in.

OUTDOOR OUTLETS AND SWITCHES

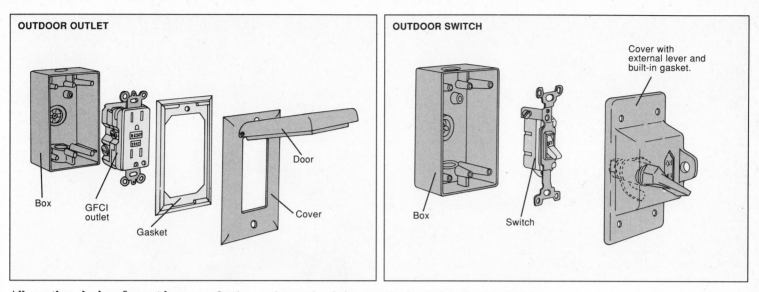

OUTDOOR OUTLET

Box

GFCI outlet

Gasket

Door

Cover

OUTDOOR SWITCH

Cover with external lever and built-in gasket.

Box

Switch

All-weather devices for outdoor use. Outdoor outlets and switches may be located on the exterior of the house or garage or in other areas of the yard. These devices must be housed in heavy-grade, rust-resistant boxes that protect them from the weather. For increased safety, codes now require that all outdoor outlets be protected by a GFCI (ground-fault circuit interrupter), which trips within 1/40 of a second, cutting off power to the circuit to protect against accidental electrical shock *(page 25)*. If the outdoor circuit is protected by an indoor GFCI, a regular outlet can be used outdoors. If the circuit is not protected, an outdoor GFCI is required.

Both the regular outdoor outlet and the GFCI outlet are made weatherproof by heavy covers and spring-loaded doors. The door on the GFCI outlet *(above, left)* opens to allow access to the receptacles and the test and reset buttons. These covers provide weatherproof protection only when closed. If a plug is to be left in the outlet for long periods—as for a low-voltage lighting transformer—the outlet must be installed where it will be protected from weather.

The outdoor switch has a cover with an external lever that flips the switch on or off without having to open the box *(above, right)*, or a cover with a small spring-loaded door that opens to expose the toggle.

SERVICING OUTDOOR DEVICES

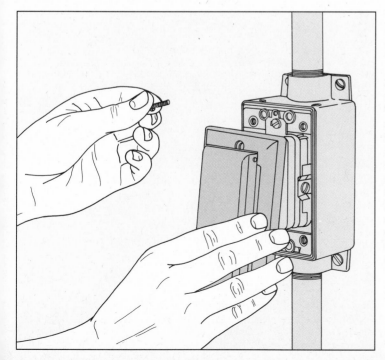

Inspecting covers and connections. Regular maintenance can prolong the life of outdoor outlets and switches and ensure their safe operation. Turn off power to the circuit by removing the fuse or tripping the circuit breaker *(page 18)*. Use a voltage tester to confirm that the power is off, following the procedure for indoor outlets *(page 86)* and switches *(page 70)*. Once the power is confirmed off, remove the outlet or switch, wipe away any moisture inside the box and inspect it. Check that the grounding wire is attached securely to the back of the box, and tighten the grounding screw. Look for dirty or corroded connections; clean the wires and screw terminals and clip back the wire ends if necessary *(page 141)*. Tighten the connections at the terminals and wire caps. Replace faulty outdoor devices as you would indoor outlets and switches. Check the seals and gaskets and apply a small amount of penetrating oil to keep them soft and pliable. If they are cracked or brittle, replace them. To make a new gasket, cut to size a rubberized cork gasket sheet purchased at an auto-supply dealer. Mount the cover so that it fits squarely against the box, as shown. Apply silicone sealant to the screw holes and around the joints between the conduit and box.

TOOLS & TECHNIQUES

This section introduces basic tests and repairs for home electrical systems, from checking for voltage to working with wiring. You can handle most electrical problems with the basic tools shown below. Although some—particularly the pliers—may resemble items found in an all-purpose tool kit, these tools are designed to ensure the safety, neatness and precision that electrical jobs require. Specialized tools, such as a multitester, can sometimes be rented from the same electrical or lighting supplier where you buy replacement parts.

For the best results, buy the best tools you can afford, use the right tools for the job, and take the time to care for and

store them properly. Clean metal tools with a rag moistened with a few drops of light oil (but don't oil the handles). To remove rust, rub with fine steel wool or emery cloth. Protect tools in a sturdy plastic or metal toolbox, with a secure lock if stored around children.

To prevent accidents, keep your work area well-lit, clean and free of clutter. Wear heavy rubber gloves when working at the main service panel or on a live circuit, and safety goggles when hammering or chiseling. Wear rubber boots when working on outdoor fixtures or in a damp basement. Use only pliers and screwdrivers with insulated handles.

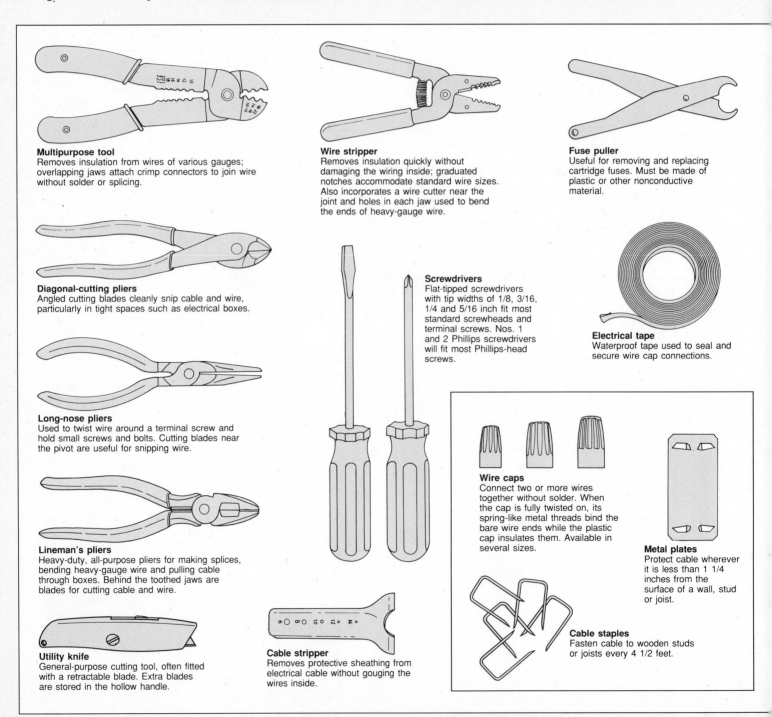

Multipurpose tool
Removes insulation from wires of various gauges; overlapping jaws attach crimp connectors to join wire without solder or splicing.

Diagonal-cutting pliers
Angled cutting blades cleanly snip cable and wire, particularly in tight spaces such as electrical boxes.

Long-nose pliers
Used to twist wire around a terminal screw and hold small screws and bolts. Cutting blades near the pivot are useful for snipping wire.

Lineman's pliers
Heavy-duty, all-purpose pliers for making splices, bending heavy-gauge wire and pulling cable through boxes. Behind the toothed jaws are blades for cutting cable and wire.

Utility knife
General-purpose cutting tool, often fitted with a retractable blade. Extra blades are stored in the hollow handle.

Cable stripper
Removes protective sheathing from electrical cable without gouging the wires inside.

Wire stripper
Removes insulation quickly without damaging the wiring inside; graduated notches accommodate standard wire sizes. Also incorporates a wire cutter near the joint and holes in each jaw used to bend the ends of heavy-gauge wire.

Screwdrivers
Flat-tipped screwdrivers with tip widths of 1/8, 3/16, 1/4 and 5/16 inch fit most standard screwheads and terminal screws. Nos. 1 and 2 Phillips screwdrivers will fit most Phillips-head screws.

Fuse puller
Useful for removing and replacing cartridge fuses. Must be made of plastic or other nonconductive material.

Electrical tape
Waterproof tape used to seal and secure wire cap connections.

Wire caps
Connect two or more wires together without solder. When the cap is fully twisted on, its spring-like metal threads bind the bare wire ends while the plastic cap insulates them. Available in several sizes.

Metal plates
Protect cable wherever it is less than 1 1/4 inches from the surface of a wall, stud or joist.

Cable staples
Fasten cable to wooden studs or joists every 4 1/2 feet.

Throughout this book the inexpensive but indispensable voltage tester is used to reveal whether current is flowing between two points of a circuit. As a safety precaution in almost every repair, you will be directed to turn off power to the circuit at the service panel, then confirm that the power is off by checking the switch, outlet or lighting fixture with the voltage tester.

Two other simple testing devices—the continuity tester and multitester—are used to determine whether an electrical part is doing its job. Both testers are battery-powered and send a small electrical current through the part being tested. The continuity tester simply indicates whether or not the current is passing through the part. The multitester, also called a volt-ohmmeter, provides a precise measurement of the strength of the current and any resistance it encounters in the circuit. The receptacle analyzer, used for diagnosing wiring problems in an outlet box, is a handy addition to the electrical tool kit.

In electrical work, it is important to make safe and reliable wire connections, and to double-check your work. Whether wire ends are joined under a wire cap or looped around a terminal screw, the connection must be mechanically and electrically secure so that current will flow smoothly from one part of the circuit to another.

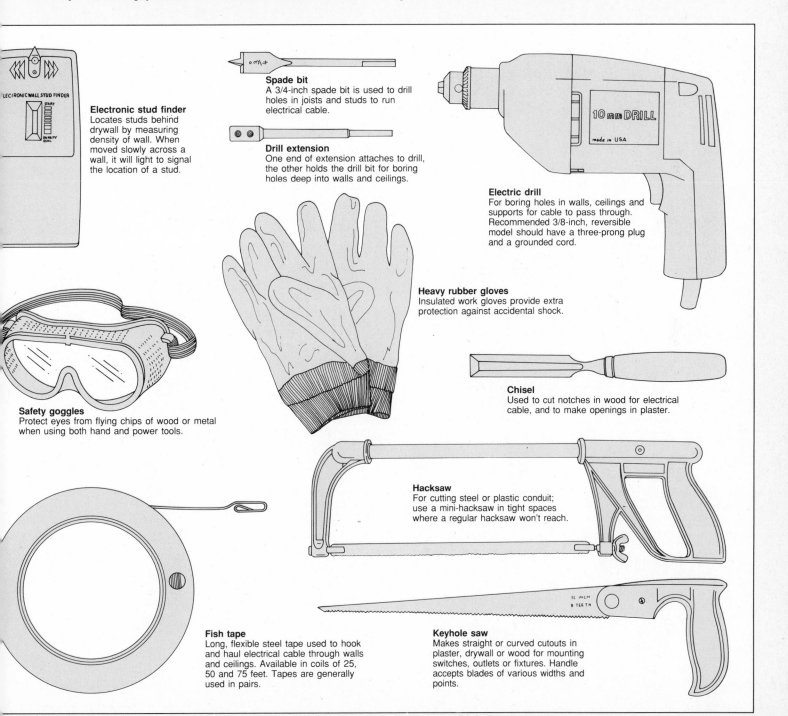

Electronic stud finder
Locates studs behind drywall by measuring density of wall. When moved slowly across a wall, it will light to signal the location of a stud.

Spade bit
A 3/4-inch spade bit is used to drill holes in joists and studs to run electrical cable.

Drill extension
One end of extension attaches to drill, the other holds the drill bit for boring holes deep into walls and ceilings.

Electric drill
For boring holes in walls, ceilings and supports for cable to pass through. Recommended 3/8-inch, reversible model should have a three-prong plug and a grounded cord.

Heavy rubber gloves
Insulated work gloves provide extra protection against accidental shock.

Chisel
Used to cut notches in wood for electrical cable, and to make openings in plaster.

Safety goggles
Protect eyes from flying chips of wood or metal when using both hand and power tools.

Hacksaw
For cutting steel or plastic conduit; use a mini-hacksaw in tight spaces where a regular hacksaw won't reach.

Fish tape
Long, flexible steel tape used to hook and haul electrical cable through walls and ceilings. Available in coils of 25, 50 and 75 feet. Tapes are generally used in pairs.

Keyhole saw
Makes straight or curved cutouts in plaster, drywall or wood for mounting switches, outlets or fixtures. Handle accepts blades of various widths and points.

DIAGNOSING ELECTRICAL PROBLEMS

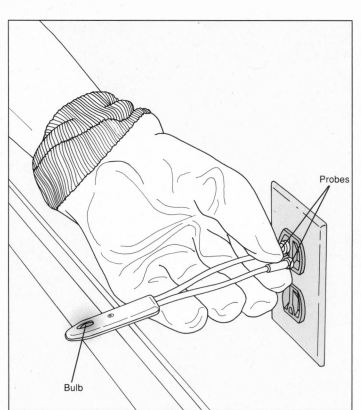

Probes

Bulb

Testing for voltage. Use a voltage tester *(left)* to make sure that the electricity to a circuit has been turned off before you work on it. The voltage tester has no power source of its own; it lights when its probes are touched to terminals or wires or inserted in slots that are charged with electricity. The voltage tester is primarily a safeguard against shock, used to make sure that there is no current in the circuit you are working on.

In some cases the voltage tester is used with the power on to locate a hot (current-carrying) wire or to test the grounding at an outlet. Work with care when performing a live voltage test. Hold the tester probes by the insulated handles or use one hand only, as shown. (For additional protection against shock, wear a heavy rubber glove.)

Buy a voltage tester rated for both 120- and 240-volt household current. Always hold the two probes by their insulation—never touch the bare metal ends. Before using the voltage tester, check to make sure it works. Plug a lamp into a working outlet. If the lamp lights, unplug it and insert the probes of the voltage tester into the slots of the outlet. If the bulb glows without flickering, the tester is good. Follow the specific directions in each chapter to turn off power to a circuit, then confirm that it is off using the voltage tester.

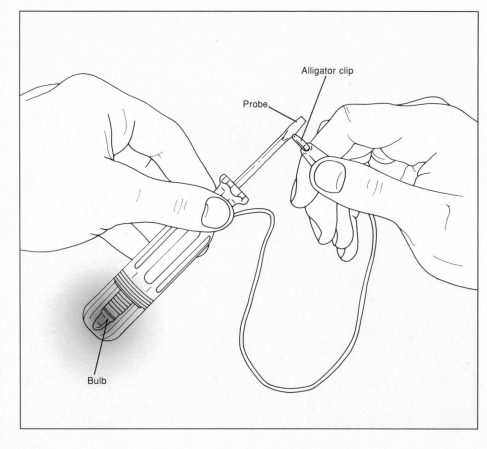

Alligator clip

Probe

Bulb

Using a continuity tester. The continuity tester sends a small current through a circuit to determine if its electrical path is intact. Because it is battery-powered, the tester must be used only when power to a circuit has been turned off. First check the battery by touching the alligator clip to the probe *(left)*; the bulb should light. To test for continuity, attach the alligator clip to one end of the circuit (a plug prong, for example) and touch the probe to the other end (the bare cord wire). If the circuit is complete, the bulb will glow. If it fails to light, there is a break in the circuit. When the tester is not in use, attach the alligator clip to its plastic insulation to prevent it from accidentally contacting the probe and wearing out the battery.

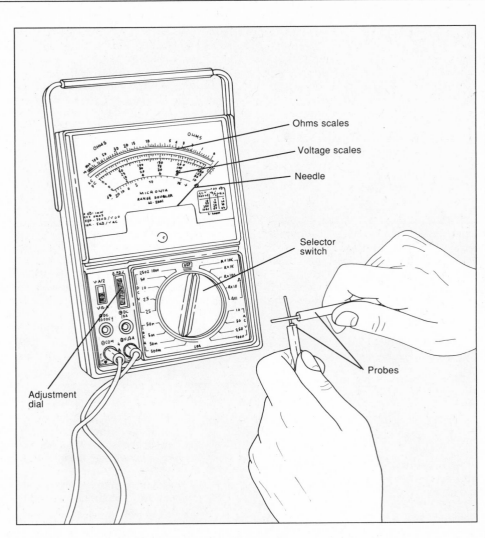

Ohms scales

Voltage scales

Needle

Selector switch

Probes

Adjustment dial

Using a multitester. A multitester gives an exact reading of the amount of resistance present in an electrical circuit (as measured in ohms), as well as high and low voltages for AC and DC current. To use the multitester, first set the selector switch on the specified scale. To ensure an accurate reading, "zero" the multitester by touching the probes together. The needle should sweep from left to right down to ZERO; turn the adjustment dial until the needle aligns directly over ZERO. To get a proper reading, the probes should firmly contact bare metal surfaces, terminals or wire ends, not insulated or dirty ones. To prevent damaging the multitester, always turn off power to the circuit when testing for resistance.

The multitester's capacity to measure low voltage makes it ideal for testing doorbells and outdoor lighting systems. Test a low-voltage circuit by setting the multitester to 50 volts on the ACV scale.

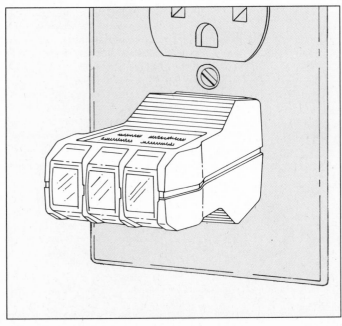

Using a receptacle analyzer. With the power on, the three probes of the receptacle analyzer fit into the slots of a grounded outlet. Three small display lights indicate whether there is current to the outlet, if the hot and neutral wires are reversed, and whether or not the outlet is properly grounded. If the analyzer reveals a wiring fault, service the outlet connections or replace the outlet *(page 86)*.

CABLES AND WIRES

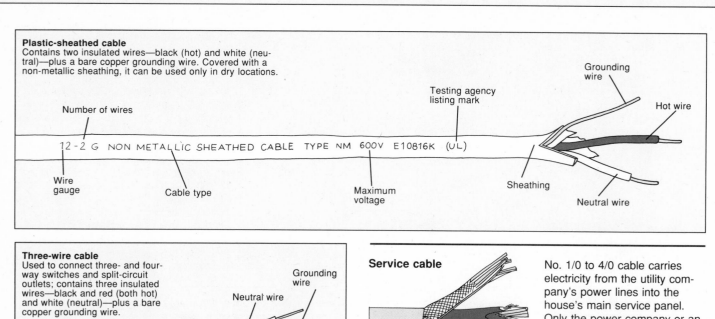

Plastic-sheathed cable
Contains two insulated wires—black (hot) and white (neutral)—plus a bare copper grounding wire. Covered with a non-metallic sheathing, it can be used only in dry locations.

Number of wires

Testing agency listing mark

Grounding wire

Hot wire

12-2 G NON METALLIC SHEATHED CABLE TYPE NM 600V E10816K (UL)

Wire gauge

Cable type

Maximum voltage

Sheathing

Neutral wire

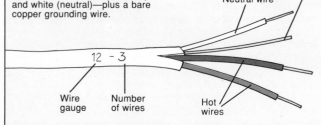

Three-wire cable
Used to connect three- and four-way switches and split-circuit outlets; contains three insulated wires—black and red (both hot) and white (neutral)—plus a bare copper grounding wire.

Grounding wire

Neutral wire

12 - 3

Wire gauge

Number of wires

Hot wires

Reading a cable. When more than one insulated wire is enclosed in a covering, it is generally referred to as a cable. Plastic-sheathed cable *(above)* is the type most often used to carry electricity throughout the home. It contains insulated black (hot) and white (neutral) wires, along with a bare copper grounding wire.

Type NM cable, known by the trade name Romex, is widely used in home wiring but must not be installed where it will be exposed to dampness. The solid plastic sheathing of type NMC cable keeps out moisture, making it suitable for use in basements or laundry rooms. Waterproof UF cable, which can be buried directly in the ground, is an excellent choice for outdoor circuits. Wiring may also be protected by flexible steel armor, commonly called BX cable.

The diameter, or thickness, of the wires in the cable is indicated by a gauge number, usually printed on the insulation. The smaller the number, the thicker the wire and the more current it can carry. The smallest wires, No. 18 and No. 16, are used to connect low-voltage systems. No. 14 or No. 12 wire is installed on most 120-volt circuits. No. 10 wire is used for dishwashers and clothes dryers; appliances that run on 240-volt current take heavy-duty No. 8 or No. 6 wire.

Be sure to match the amperage rating of any device you install with that of the wiring. Overloaded wiring can pose a fire hazard. Consult the chart below for wire amperage ratings.

WIRE GAUGE AND AMPERAGE

Wire Gauge	No. 6	No. 8	No. 10	No. 12	No. 14	No. 16	No. 18
Amperes	55	40	30	20	15	10	7

Service cable

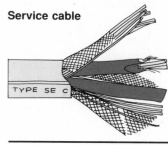

TYPE SE c

No. 1/0 to 4/0 cable carries electricity from the utility company's power lines into the house's main service panel. Only the power company or an electrician should work on this wiring.

Armored cable

Often found in older homes, armored cable contains two wires and a metal bonding strip or grounding wire wrapped in paper and sheathed in a flexible steel housing. When no grounding wire is present, the housing and bonding strip act as ground.

Zip cord

Nos. 16 and 18 wire are made of fine metal strands covered with rubber or plastic insulation and used for lamps and small appliances. Polarized cord is used where the hot and neutral wires must be aligned throughout a circuit. One of the two wires is keyed as neutral—usually with a molded ridge on its insulation.

Low-voltage wire

Low-voltage or bell wire is a single-strand No. 16 or No. 18 wire used for doorbells and patio lights. Wire ends can be joined by twisting them together and screwing on a wire cap of the appropriate size or by securing them in a crimp connector.

WORKING WITH PLASTIC-SHEATHED CABLE

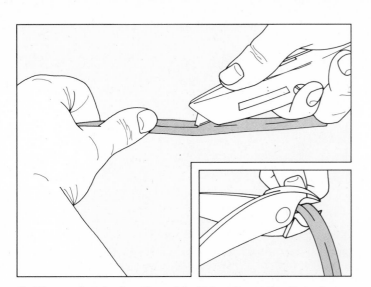

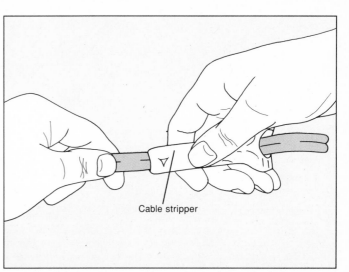

Cable stripper

Cutting and stripping the cable. Using diagonal-cutting pliers, carefully cut through the plastic sheathing 8 inches from the end of the cable *(inset)*. Next, adjust a utility knife so that only the corner of the blade protrudes from the handle. Insert the blade in the sheathing and slice toward the end of the cable, as shown. Hold the knife at an angle so you do not cut into the wiring. Peel back the sheathing to expose the wires, then trim it with diagonal-cutting pliers. To strip back the wire ends, see page 138.

Using a cable stripper. If you are cutting and stripping several cables, an inexpensive cable stripper will make the job easier. Slide the cable stripper over the sheathing until it is 8 inches from the end of the cable. Squeeze the handles together so that the blade bites into the sheathing, but be careful not to gouge the wires inside. Pull the cable stripper along the sheathing to the end of the cable. Peel back the sheathing to expose the wires, then trim it with diagonal-cutting pliers.

WORKING WITH ARMORED CABLE

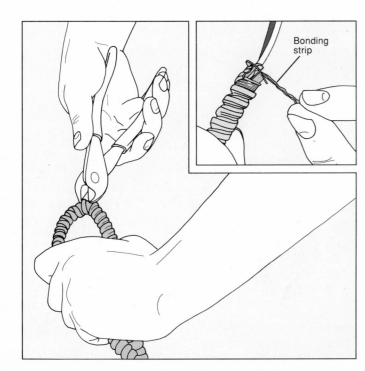

Bonding strip

Wiring protected by steel. Armored cable has two insulated wires covered with paper and encased in a spirally wound, interlocking steel jacket. Still permitted by some local codes, it can be useful in locations where the wires need additional protection. Armored cable can only be installed indoors in a dry location.

The steel sheathing of armored cable can be cut in one of two ways. If you have a reasonably strong grip, you can bend the cable in two and squeeze it until the casing separates *(left)*. Then use a pair of diagonal-cutting pliers to cut off the casing and metal grounding strip, exposing the insulated wires. Or set the cable on a firm surface and cut partway through the casing with a hacksaw at a right angle to the spirals. Twist the armor until the spirals separate.

Cap the end of the metal jacket with a plastic or fiber bushing to protect the wires. If the armored cable contains a bonding strip *(inset)*, fold it back over the steel sheathing. If the cable contains a bare grounding wire, connect it as you would for standard plastic-sheathed cable.

STRIPPING NEW WIRES

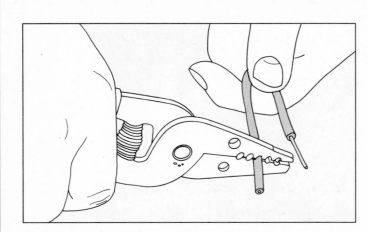

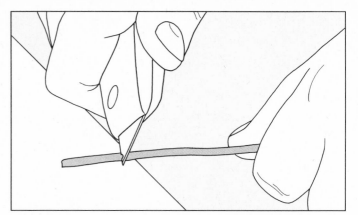

Using wire strippers. A stripping tool with a series of holes to fit various wire gauges snips through insulation without damaging the metal wire. Insert the wire into a matching slot on a pair of wire strippers *(above)* or multipurpose tool. (The gauges of the wire and slot must be the same.) Close the tool and twist it back and forth until the insulation is severed and can be pulled off the wire. Strip back 3/4 inch of insulation from the new wire to make most connections.

Using a knife. A sharp knife requires more care than wire strippers, but can also be used to strip the insulation from a wire. Place the wire on a firm surface and carefully cut into the insulation, paring it away at an angle, as shown, and taking care not to gouge the wire inside. Always cut away from you in case the knife slips. Turn the wire over and make a second cut, then pull off the insulation. If you have nicked the wire, use diagonal-cutting pliers to snip off the wire end, then start again.

MAKING TERMINAL CONNECTIONS

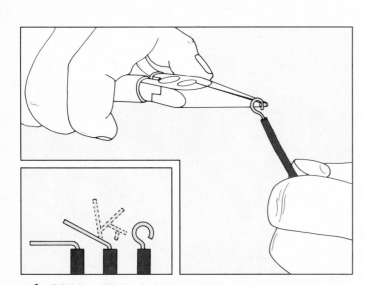

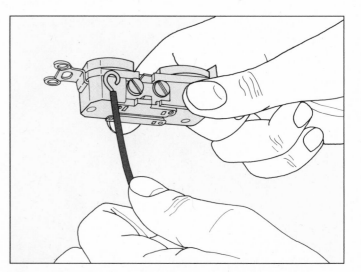

1 Making the hook. Strip back the insulation from the wire end, then bend it at a right angle with long-nose pliers *(inset)*. Starting near the insulation, make progressive bends to the right, moving the pliers toward the wire end until an open hook is formed.

2 Making the connection. Loosen the terminal screw, but do not force it out of its threads. Attach the wire end to the terminal so that its hook will close in the same clockwise direction as the screw *(above)*. Make sure that the hook loops at least three-quarters of the way around the screw. Tighten the screw so that it grips the wire end securely (it should flatten the wire slightly). If part of the bare wire end is exposed, take apart the connection and start again.

WORKING WITH STRANDED WIRE

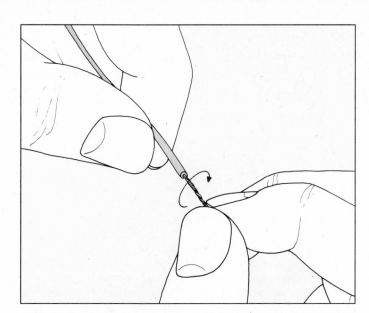

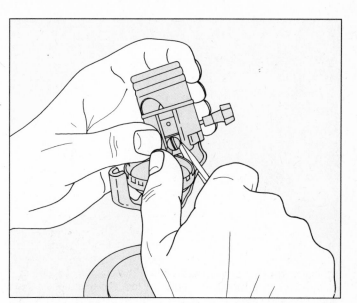

1 **Preparing the wire end.** Use wire strippers or a multipurpose tool to strip about 3/4 inch of insulation from the wire end, taking care not to nick the strands. Twist the bare strands together tightly in a clockwise direction between your thumb and forefinger, as shown. If there are any stray or broken strands, snip the wire end and start again. To attach the wire end to a terminal screw, continue to step 2. To join it to a solid wire, go to step 3.

2 **Making a terminal connection.** Form the twisted strands into a hook. Attach the wire end to the terminal so that the hook will close in the same clockwise direction as the screw. Use your thumb and the tip of a screwdriver to force the twisted strands under the terminal *(above)*, then tighten the screw so that it flattens the wire slightly. If part of the bare wire end is exposed, or there are any stray strands, take apart the connection and start again.

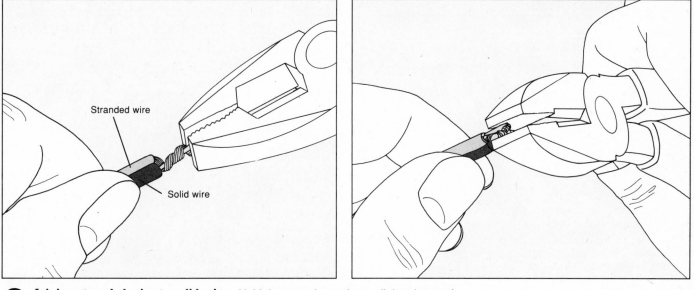

Stranded wire

Solid wire

3 **Joining stranded wire to solid wire.** Hold the two wire ends parallel and wrap the stranded wire in a spiral around the solid wire *(above, left)*. With lineman's pliers, fold the end of the solid wire over the wrapped portion *(above, right)*, then screw on a wire cap to secure the connection. To join stranded wire to stranded wire, hold the wire ends parallel and twist them together tightly in a clockwise direction.

PUSH-IN TERMINALS

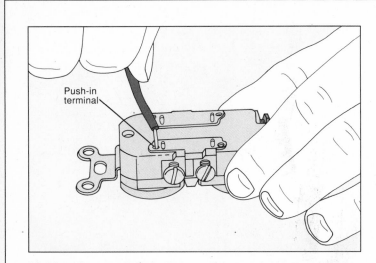

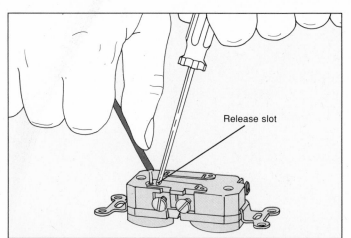

Making the connection. Some switches and outlets are fitted with push-in terminals that automatically grip the wire ends. This feature eliminates the need to bend the wire ends into a hook, but does not provide as secure a connection as screw terminals. To attach a wire to a push-in terminal, strip back the insulation according to the strip gauge marked on the device—usually 1/2 inch—and push the bare wire end into the hole up to its insulation. Make sure that no uninsulated wire is exposed. Note that aluminum wire should never be used with push-in terminals.

Releasing a push-in connection. To free the wire from a push-in terminal, insert a small screwdriver or stiff wire into the release slot next to the terminal hole, then pull on the wire. If the wire end breaks off in the terminal, replace the switch or outlet.

WIRE CAP CONNECTIONS

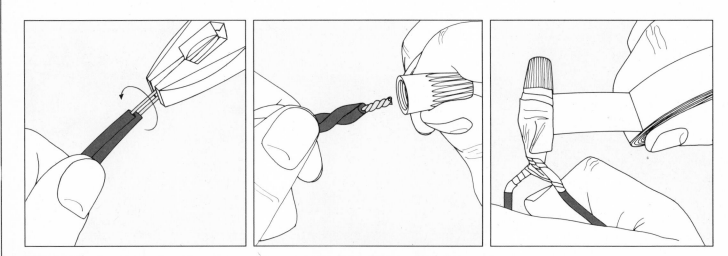

Using wire caps. Using wire strippers or a multipurpose tool, remove the insulation from each wire end, taking care not to damage the wire inside. Hold the wires side by side with one hand, grip the bare ends with pliers *(above, left)* and twist them together clockwise until the turns are tight and uniform along the entire length of the bare wires. Slip a wire cap over the connection *(above, center)* and screw the cap clockwise until it is tight and no bare wire remains exposed. Test the connection with a slight tug. To secure the connection, wrap electrical tape around the base of the cap, then once or twice around the wires, and finally around the base of the cap again *(above, right)*.

JUMPERS AND PIGTAILS

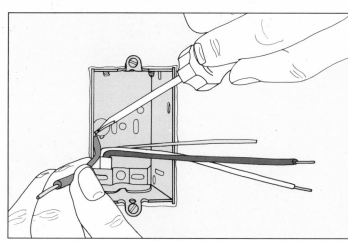

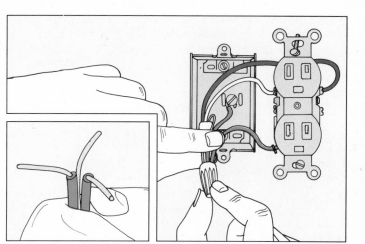

Joining jumper wires in a pigtail connection. When there is more than one wire in the box intended for the same terminal, use a jumper wire to make the connection. A jumper is usually 6 inches long and must match the gauge of the wires it joins. When grounding an outlet with one cable in the box, for example, you will need two jumpers—one to attach to the grounding terminal on the outlet, and the other to attach to the grounding screw at the back of the box. Cut the first wire, strip back both ends *(page 138)*, then form one end into a hook and attach it to the grounding

screw at the back of the box *(above, left)*. Prepare the second jumper in the same manner and attach it to the grounding terminal on the outlet. Join the jumpers and the bare copper grounding wire using a pigtail connection. Hold the wire ends together and bend each at a right angle *(inset)*. With lineman's pliers, grip the wire ends and twist them together in a clockwise direction. Continue twisting until the turns are tight and uniform along the entire length of the bare wires. Clip 1/4 inch off the end of the twisted wires, then screw a wire cap onto the connection *(above, right)*.

SERVICING WIRES IN A BOX

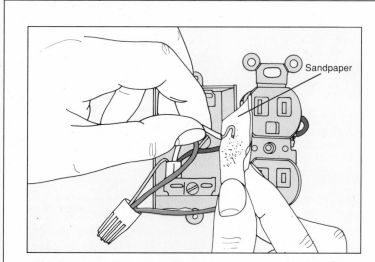

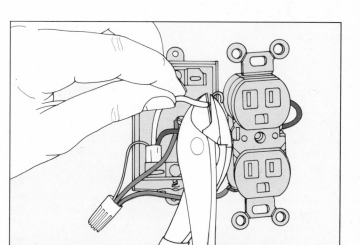

1 Cleaning the connections. Dirty or loose connections in an electrical box can produce sparks or shocks. Once you have confirmed that the power is off, remove the device from the box and examine the wire connections. If they appear dirty or corroded, disconnect them, then use fine sandpaper to burnish them *(above)*. If they are discolored, cleaning won't help. Blackened wire should be clipped back and restripped *(next step)*; a switch or outlet with blackened terminals should be replaced.

2 Clipping back damaged wire ends. Use diagonal-cutting pliers to snip off a damaged wire end at the edge of the insulation, as shown. Use wire strippers to strip 3/4 inch of insulation, exposing a clean wire end. Next, shape the wire end into a hook *(page 138)* and curl it around its terminal, or retwist the wire connection *(page 140)*.

INDEX

Page references in *italics* indicate an illustration of the subject mentioned. Page references in **bold** indicate a Troubleshooting Guide for the subject mentioned.

A

Adapter plugs, *89*, *99*
Aluminum wiring, 20
Amperage, 15, 23

B

Boxes. *See* Electrical boxes
Bulb-life extenders, 28

C

Cables. *See* Wires
Cable strippers, *132*, *137*
Ceiling fans, **56**, 43, *56-57*
Ceiling fixtures, **44**, *45-47*
    Chandeliers, *49, 51*
    Recessed, *54-55*
    Track, *52-53*
    Wiring, *51, 110-111, 117*
Chandeliers, **44**, *49-51*
Circuit breakers, *10*, 15, *16, 18*
Circuits. *See* House circuits
Conductors, 15
Continuity testers, 133, *134*
Cords:
    Lamps, 26-27, 35
    Maintenance, *17*
    Outdoor, 126, *128*
    Round, 35

D

Dimmer switches, *35, 79*
Doorbells, **66**, *67*
    Bell or chime unit, *69*
    Push button, *67-68*
    Transformers, 66, *68*

E

Electrical boxes, *14*, 15, 102-103, *114, 118*
    Circuit extensions, *118-119*
    Installation, *115-118*
    Outlets, *86*
    Repairs within, *104, 141*
    Switches, *73*
    See also Grounding

Electrical codes, 14
    See also Electrical specifications
Electrical current, 15
Electrical shocks, 8, *10*
Electrical specifications:
    Amperage, 15
    Fluorescent tubes, *60*
    Incandescent bulbs, *28*
    Outlets, *86*
    Switches, *72*
    Voltage, 15
    Wattage, 15
    Wires, *136*
Electricians. *See* Service calls
Electric ranges, 86, *99-100*
Emergency procedures, **9**, 8
    Electric shocks, 8, *10*
    Fire, 8, *11, 12*
    Power failures, *13*
    Sparks, *12*, 16
Extension cords, 17, *89*

F

Fire extinguishers, *11, 12*
Fires:
    Safety procedures, 8, *11, 12*
Fish tapes, 103, *133*
    See also House circuits: Fishing
Fluorescent lighting, **59**, 58
    Ballasts, 58, *63, 64, 65*
    Fixture replacement, *63-64*
    Grounding, *62*
    Lamps, 58
        one-socket, *65*
        trigger switch, *64*
    Rapid-start, *58, 59, 62*
    Sockets, 58, *62*
    Starter type, 58, *59, 61*
    Tubes, *60, 61*
Fuses, *10*, 15, 16, *18-19*

G

Generators, portable, *13*
Ground-fault circuit interrupters, 16, 17, 86, *89*, *96-97*, 98
Grounding, 15, *24*, 118
    Adapter plugs, *89, 99*
    Isolated grounds, *88*
    Outlets, 86, *89*, *98-99*
    Rapid-start fluorescent lights, *62*
    Wall switches, *73*, 118
    Wires, *14*

H

Halogen lighting, 27, *42*
Hot wires, 15, *136*
House circuits, **103**, 14, *102-103*
    Dedicated, 15
    Extensions, *102-103*
        through attics, *107-108*
        through baseboards, *109*
        through basements, *105-106*
        through ceilings, *110-111*
        through walls, 102, *110*
    Fishing, 103, *105-106, 107, 108, 109*
    Inspections, *20*
    Mapping, *20-22*
    120-volt, 15
    Raceway cables, 103, *112-113*
    Safety procedures, 103
    240-volt, 15
    See also Electrical boxes; Outlets; Switches
Household appliances, 23

I-J

Incandescent bulbs, *28, 29*
Insulators, 15
Jumpers, *20, 141*

L

Lamps, **27**, *26*
    Bulbs, *28, 29*
    Cords, 26-27, 35
    Fluorescent, 58, *64-65*
    Halogen, 27, *42*
    Plugs, 26, 27, *35-37*
    Rewiring, 26, *38-41*
    Shades, *30*
    Sockets, 29, *30-32*
    Switches, *32-35*
Light bulbs:
    Fluorescent tubes, *60, 61*
    Halogen, 42
    Incandescent, *28, 29*
Lighting fixtures, **44**, 43
    Ceiling, **44**, *45-47*,
        chandeliers, **44**, *49-51*
        recessed, *54-55*
        track, *52-53*
    Installation, *118*
    Polarization, *25*, 43
    Safety procedures, 43
    Wall, *48-49*
    See also Fluorescent lighting; Lamps; Outdoor lighting
Load calculation, 23

ACKNOWLEDGMENTS

The editors wish to thank the following:
Richard Day, Palomar Mountain, Calif.; Vince Di Cesare, Hubbell Canada, Inc., St. Laurent, Que.; Dickie Moore Rentals, St. Laurent, Que.; Bruce Emerson, Emerson Electrical Supply, Salisbury, Mass.; Leo Frenette, H.R. Cassidy Ltd., Montreal, Que.; Roy L. Hicks, Ontario Hydro, Toronto, Ont.; Claude Lalande, Montreal, Que.; Orf Lalli, Lightolier Inc., Lachine, Que.; Henri P. Laporte, Philips Electronique Ltée., St. Jérôme, Que.; Kenneth Larsen, C. Howard Simpkin Ltd., Montreal, Que.; Judy Mann, New Haven, Conn.; Henry R. Martinez, Leviton Manufacturing Co., Inc., Gardena, Calif.; David J. Morrissey, The Wiremold Company, West Hartford, Conn.; Nutone Inc., Cincinnati, Ohio; Mort Schultz, Plantation, Fla.; Steven A. Schmit, Underwriters' Laboratories Inc., Northbrook, Ill.; David E. Soffrin, Edison Electric Institute, Washington, D.C.; Joseph E. Stine, Lee Electric Co., Baltimore, Md.; Seng-Tee Tan, Loran, Inc., Redlands, Calif.; Union Electric Supply Co., Ltd., Montreal, Que.; Norman Zimmerman, Universal Ship Supply, Ltd., Montreal, Que.

The following persons also assisted in the preparation of this book:
Mary Ashley, René Bertrand, Philippe Arnoldi, Diane Denoncourt, Michael Kleiza, Lauren E. Reid, Odette Sévigny, Natalie Watanabe, Billy Wisse.

Typeset on Texet Live Image Publishing System.